# Naval Aviation Excellence

David Maybury

Naval Aviation Excellence by David Maybury

Copyright © David Maybury, 2023

**Notes to Readers:** A more narrative version of this material, including additional stories and reflections is available in *Inside Naval Aviation: A Memoir*. It is available as an audiobook, ebook or in print. The audiobook is available on Audible, Spotify and other major platforms.

When you finish the book, a short, written review would be greatly appreciated to help others discover it.

Front Cover: Lieutenant Frank Morley on approach for the very first Super Hornet landing aboard an aircraft carrier, the USS Stennis. Navy flight test photographer, Randy 'Fireball' Hepp, took the image while his Navy pilot, Tom Gurney, flew close formation.

Back Cover: Four VA-115 Eagle A-6E Intruders flying past Mount Fuji close to their home base, NAF Atsugi, near Tokyo, Japan. Photo by Eagle BN, Greg 'Nubian' Eaton.

A glossary of acronyms is provided at the back of the book.

Paperback ISBN: 979-8-9894359-9-9
Ebook ISBN:      979-8-9894359-4-4
Hardcover ISBN: 979-8-9894359-0-6

Table of Contents

Preface

# Acknowledgements

I have been blessed with amazing supporters. Key individuals include those below.

My parents, Paul and Rose, siblings Janet, Joan, Anita and Glen, high school and university teachers, friends Clay, Steve, Bill, Neal, and Bobby, classmates and roommates Jim, Tom, TC, Sam, Joe and Paul.

My ex-wife, Wendy, for sharing exciting adventures including the birth of our beautiful daughter, Elizabeth. My wife, Michelle, Elizabeth, sons Cameron, Jarred and Ethan have always been a source of love and encouragement. Michelle has provided unending patience with repeatedly listening to stories then during the first COVID years while they were written down.

A-7 pilot, Commander Ed Ohlert, brought the excitement of naval aviation to the midshipman at the University of Colorado. With a twinkle in his eyes and a broad smile, he shared carrier stories that established a foundation of eager anticipation. He inspired this book.

Lyle Mudge retired from Boeing after filling the roles of Engineer, Team Leader, Project Manager and others while I've known him. He has been a good friend providing solid counsel since 1993.

Steve Donaldson has superbly served the Navy in various capacities while at NAVAIR. I am blessed by his friendship.

I am indebted to a great number of outstanding pilots, Naval Flight Officers (NFOs), engineers, maintenance and support personnel throughout my career in the United States Navy (USN). Many are mentioned by name or callsign to highlight their professionalism.

My test pilot school classmates, the instructors at U.S. Naval Test Pilot School (USNTPS), the squadron personnel of VA-115, VA-128, VX-5 and Strike were among the best in the fleet and serving with them was priceless.

Finally, a big thank you to the staff of NAPRA and our detachment in Okinawa for keeping Naval and United States Marine Corps (USMC) forward deployed aircraft flying around the globe.

This book could not have been completed without the challenging, yet amazing course of instruction taught at the USNTPS at NAS Patuxent River (PAX), Maryland[1]. While most people think of flying amazing aircraft when considering the role of test pilots, USNTPS instructs students in how to research then write test plans and reports as well as flight test techniques for new aircraft, weapons and systems. The sheer magnitude of written assignments developed an inner focus and endurance to write for extended periods of time.

Finally, it is important to recognize the critical contributions of all the spouses and family members of our active-duty military personnel, government employees and defense contractors. Mission requirements often demand extended time away from home. Carrier battle groups regularly deploy for six months. While on shore duty, extensive time away is still required.

All that time separated from one another adds stress to families. Those family members that stay at home become self-sufficient and independent out of necessity. Their faithful contributions are invaluable to their deployed and traveling spouses and ultimately, their country. We thank them for their love and support.

---

[1] Naval Air Station (NAS) Patuxent River is the home of naval aviation's developmental flight test teams including engineers, maintenance and support personnel along with test pilots.

# Preface

This memoir shares a global journey in excellence. It highlights individuals and organizations while working with NASA, aircraft carrier-based A-6E Intruder aircraft, test pilots flying supersonic F/A-18E/F Super Hornet and F-14 Tomcat aircraft, government and contractor employees and a host of others that serve to defend freedom twenty-four hours a day. Maintaining global peace and freedom is not easy but well worth the investment.

Most members of the military would never venture to write about their service in any great length. Not that the experiences weren't noteworthy, but each of us feels that our part was so small as to be insignificant. To write about our role would be egotistical and frowned upon by our comrades. Yet it is in this silence that we find ourselves misunderstood by so many people globally including our own countrymen. It is important to help clarify the perspective from those who served in the military. My experience in no way compares to those of war veterans. They deserve great admiration for conquering incredible challenges leading to freedom against tyranny and oppression.

Unfortunately, there are groups globally that develop plans to harm and kill others to advance their personal agendas. History recalls many such groups including recent, horrifying examples. Those that choose to kill and destroy often find resistance from others including peace-loving nations of the world. The United States, for example, currently uses cruise missiles, drones and manned aircraft to track and destroy murderous groups that prefer to spread worldwide terror. Manned aircraft operate with strict rules of engagement to minimize collateral damage and loss of life. I flew armed A-6E Intruder aircraft in that role.

The role of the Intruders was to engage and destroy the enemy with bombs, rockets, missiles and mines. The aircraft, crews and ordnance men were certified to load, carry and deliver nuclear weapons, if required. Our orders were clear and they enforced the national strategy of the United States government. Civilian and military leaders generated the strategy in Washington, D.C. after careful consideration of the best methods to minimize the loss of innocent life. Our job was to deliver that policy decision in the safest yet most effective manner possible.[2] One of my greatest concerns was to deliver our weapons on the correct target and avoid any civilian casualties. We, as an Intruder community, made great efforts to ensure that ordnance was delivered on target, at the time designated in our mission planning.

Many people may not appreciate that most members in the military do not seek battle or war at all[3]. They hope to bring stability and peace to an unsettled world. The great majority of men and women I served with had no passion for killing or harming others[4]. However, we were very well trained and disciplined in our work. We would attack and kill if directed to do so by our superiors. There was no hatred in our hearts for Russians, Iraqis, Afghanis, or any other people. The only time I ever saw anger in the military was for those terrorists that facilitated the 11 September attacks in the USA. Even then it wasn't hatred, but a firm resolve to eliminate the source of evil that devastated so many countries worldwide with the loss of their citizens. I personally never killed anyone and didn't destroy anything but practice targets. I never actively participated in combat but rather trained for it. Our efforts and training prepared us for war resulting in a safer world.

The United States invested considerably to cease the hostilities of WWII. President Kennedy later defined our role as a nation when he stated, "Let every nation know, whether it wishes us well or ill, that we shall pay any price, bear any burden, meet any hardship, support any friend, oppose any foe to assure the survival and the success of liberty". When speaking of our adversaries, he continued, "We dare not tempt them with weakness. For only when our arms are sufficient beyond doubt can we be certain beyond doubt that they will never be employed".[5] The modern American military was founded on these words.

---

[2] General Schwarzkoff stated onboard the USS Guam on the way to Grenada, "I asked myself why on earth the U.S. was getting involved in Grenada". Then I said, "Schwarzkoff, just let it sort itself out. You're an instrument of policy. You don't make policy". Pyle, 1991

[3] "While most Americans want peace, no one wants it more passionately than military professionals." General Colin Powell, Proceedings, May 1990.

[4] General Schwarzkoff states 'War is profanity. It really is. It's terrifying. Nobody is more anti-war than an intelligent person who's been to war.' Pyle, 1991

[5] https://www.jfklibrary.org/archives/other-resources/john-f-kennedy-speeches/inaugural-address-19610120

The Cold War with the Soviet Union dominated my first ten years of military service. Weapons were not fired despite being aimed at one another. It seemed like a cat and mouse game for those of us deployed overseas. Our military demonstrated strength to convince the Soviets and others that it was not in their interest to initiate or continue conflict.

A Soviet fighter pilot, Alex Zuyev, helped inspire me to write this book. He piloted a stolen MIG-29 fighter aircraft from the Soviet Union to Turkey in 1989. Once in Turkey, Alex requested asylum in the United States of America. A year later, he gave a Soviet air tactics briefing at my squadron. Months later, he asked me to review and comment on an early draft of his book, Fulcrum[6]. In writing his story, Alex shared very human challenges along his journey. That is often missing in military stories yet inspiring to read.

In this book, I've followed Alex's lead and shared my journey through naval aviation. Details like studying, physical fitness and social interaction are included as they were all critical to meet the demands of the profession. Making split-second decisions during physically demanding flying required knowledge, strength and endurance along with positive interactions with other crewmembers. The narrative encompasses the start of my naval career through to my retirement after twenty years of active-duty service travelling around the world.

While fortunate to fly in amazing aircraft, I intentionally haven't written many technical details to simplify the narrative. There are many others that have written excellent aircraft descriptions and I defer to them. Rather, I've recorded stories of life around the aircraft, personnel and organizations. I've been fortunate to work with gifted individuals and witness inspiring scenes from the ground and air. I was honored to play a very small part to help advance global stability and peace. My wish is for a more dramatic advancement of world peace that we have so earnestly pursued for countless years.

---

[6] A. Zuyev & M. McConnell, *Fulcrum; A Top Gun Pilot's Escape from the Soviet Empire*, (New York, NY: Warner Books, 1992)

To start the journey, words from George Will set the scene[7];

"Most of us, most of the time, live in blissful ignorance what a small elite, heroic group of Americans are doing for us night and day. As we speak, all over the globe, American sailors and submariners and aviators are doing something very dangerous. People say, "Well, it can't be too dangerous because there are no wrecks." But the reasons we don't have more accidents is that these are superb professionals: the fact that they master the dangers does not mean that the dangers aren't real. Right now, somewhere around the world, young men are landing high-performance jet aircraft on the pitching decks of aircraft carriers - at night! You can't pay people to do that: they do it out of a love of country, of adventure, of the challenge. We all benefit from it, and the very fact that we don't have to think about it tells you how superbly they're doing their job - living on the edge of danger so the rest of us need not think about, let alone experience, danger".

---

[7] George Will commenting during the ABC news special coverage on January 28, 1986, of the space shuttle Challenger disaster.

# Chapter 1

## University of Colorado (CU), Aerospace Engineering, The Navy

*The United States Navy; not just a job, but a truly amazing adventure!*

I was accepted to study aerospace engineering at the University of Colorado in Boulder commencing in August 1978. I arrived not knowing what to expect but was about to gain a priceless education in how the world designs and manufactures new technology.

I saw amazing videos of jet aircraft landing aboard aircraft carriers. It was a career path for United States Navy (USN) midshipman who made it through flight school after university graduation. I agreed to give the Navy Reserve Officers Training Corps (NROTC) a no-obligation try while applying for a scholarship.

Every summer, naval midshipmen from across the United States of America (USA) travel on a one-month summer cruise. My first cruise was to Naval Base Bremerton near Seattle aboard the USS Sacramento, AOE-1[8]. The ship was an oiler and supply replenishment ship that provided support for other ships including aircraft carriers while steaming at sea. For our second cruise, midshipmen were courted by four sections of the naval service as they tried to sell us on their specific career field. The first week was with the submarine fleet in Hawaii followed by the surface and aviation fleets in San Diego then finishing with the Marine Corps at Camp Pendleton. During my third and final cruise, I was stationed aboard the USS Benjamin Stoddert in Pearl Harbor.

Around the end of my 3rd year of Uni, several friends explained how we could get both bachelor's and master's degrees in aerospace engineering within 5 years if we attended summer school and took extra courses. I joined the group and enrolled in summer classes. We supported each other through tough academics and graduated with the two degrees after five years of study. The increased workload to complete the

---

[8] The replenishment ship Sacramento was an invaluable asset; https://www.navysite.de/ships/aoe1.htm

master's degree was challenging but I felt it would be worthwhile when complete.

After our university graduation ceremony, I joined several other first-class midshipmen as we were commissioned ensigns in the USN. I selected Air Force Major Rowland Worrell[9] to swear me into the military. He was studying for a master's degree in aerospace engineering as well. Rowland was a highly decorated aviator from the Vietnam war, an enthusiastic and positive mentor that provided invaluable support and guidance.

After graduation, I needed to wait seven months for the commencement of flight school. My midshipman counselor, Commander (CDR) Ohlert, spoke to my detailer in Washington, D.C. (the officer who assigns positions resulting in military orders). The detailer offered a position at the NASA Johnson Space Center[10], south of Houston, Texas.

Commissioning photo after university graduation in May 1983.

---

[9] Major Worrell; http://veterantributes.org/TributeDetail.php?recordID=1254
[10] National Aeronautics and Space Administration (NASA); https://www.nasa.gov/centers/johnson/home/index.html

# Chapter 2

NASA, Johnson Space Center and Ellington AFB, TX

*"You don't want to attend the easiest training but rather the most difficult. The better your training, the easier it will be to conquer whatever challenges are thrown your way".*
CDR Dick Richards, NASA astronaut

I drove from Denver to Clear Lake, Texas, south of Houston, in June 1983. I checked in with the military liaison officer at the Johnson Space Center (JSC). The NASA staff were exceptional. The outdoor rocket museum at the JSC had a Saturn Five lying horizontally allowing a close inspection of the design. This was the massive rocket that took men to the moon and was very impressive.

Three Naval Academy ensigns and I were given temporary orders to NASA awaiting flight school in Pensacola, Florida starting in January 1984. Morri and I were naval pilot candidates. Nick[11] and Greg[12] were naval flight officer candidates.

Greg and Nick were stationed at the JSC completing data analysis. Morri and I were stationed at Ellington Air Force Base (AFB), a 20-minute drive from the Johnson Space Center. Ellington AFB had several aircraft that supported the NASA mission. The astronauts flew the supersonic-capable, 2-seat, T-38 trainer aircraft, a KC-135 Zero 'G' aircraft affectionately named the Vomit Comet and the Shuttle Training Aircraft (STA).

Morri was assigned a crew position for the KC-135 Vomit Comet. He came to work in a flight suit, flew on all KC-135 missions, logged a lot of flight time and met many astronauts, engineers and their experiments. The KC-135 flew parabolic profiles to provide astronauts and scientists zero-G time like experienced in space. The aircraft earned its name the Vomit Comet as most people got very sick on their first flight. Many people got sick every time they flew it.

---

[11] Nick and I worked together on Hornet engineering projects in Washington, D.C. in the 1993-1994 timeframe.
[12] Greg was in the class ahead of me at Naval Test Pilot School in 1989.

On typical missions, the KC-135 pilot flew the aircraft up to 40,000 then brought the power to idle while pushing forward on the control yoke to generate a weightless free fall back to Earth. The free fall lasted approximately 30 seconds after which the pilot applied back pressure on the yoke and recovered the aircraft using a 2 G pullup maneuver to regain level flight. Once level, the pilot reapplied power to climb back up to 40,000 and repeat the process over again. A typical mission had the KC-135 perform forty, zero-G parabolas. I was able to fly several zero missions on the Vomit Comet. I got sick on my first flight but was able to enjoy the environment on future flights. The zero-G part was great fun and I always wanted to do more.

From left; Morri, crew leader, Bob and I floating in zero G.

My duty station at NASA was with the STA team. The Space Shuttle launched from earth with solid rocket motors and liquid fired engines but returned for landing as an unpowered glider. The approach and landing profile had to be perfect as there was no power to facilitate a go-around and second attempt. The STA was a highly modified Gulfstream II business jet that could simulate the flying characteristics of the glider Space Shuttle. It had standard Gulfstream controls and displays on the right, Pilot In Command (PIC) side of the aircraft. The left pilot station was a functional replica of the Space Shuttle cockpit. In the back of the aircraft was an array of computers that facilitated a change in the flying qualities from a Gulfstream business jet into a Shuttle simulator.

The PIC took off the aircraft from the right seat with an astronaut seated in the left seat and proceeded out to a training area over the Gulf of Mexico at an altitude of 40,000 feet. The aircraft was slowed to the landing gear extension speed of approximately 250 nautical miles (nmi) per hour (Knots) Indicated Airspeed (KIAS) and the main gear were dropped. The nosegear remained stowed due to air load considerations during the simulation profiles. The Space Shuttle simulation mode was engaged driving the engines into 90% reverse thrust, a belly mounted speed brake deployed, the upper wing spoilers deployed upward to dump lift and create drag and the flaps came down to add drag and lift. These changes turned a normally sleek business jet into a flying brick like the Space Shuttle.

At this point, the astronaut in the left seat of the aircraft took control of the aircraft and flew nominal Shuttle arrival profiles to a simulated landing point. The mission computers in the back displayed all the shuttle avionics information on the multifunction displays and Heads Up Displays (HUDs) in front of both astronaut and copilot. Flying the Shuttle back for landing was like playing a video game. The STA gave the astronauts practice playing the game with the challenges of real-world clouds, visibility, wind shear and turbulence. The flight profiles were recorded with onboard equipment to facilitate feedback and postflight debrief of the astronauts on their performance. If anything were to ever go wrong during the descent, the crew had already briefed and practiced that the Space Shuttle mode would be disengaged, and the PIC would recover the aircraft using standard Gulfstream II flight controls.

NASA Shuttle Training aircraft in approach to landing in Shuttle flying qualities mode with a T-38 supersonic trainer in chase.

A Shuttle Training Aircraft with the astronaut shuttle controls on the left and standard Gulfstream controls right.

I normally came to work in my khaki uniform, worked with engineers, analyzed data, wrote computer program code to assist in data analysis and occasionally got to fly in the STA. Meeting the astronauts was common as the STA staff directly supported them. Whenever a new Shuttle software build was released, the astronauts flew it first aboard the STA in actual flight conditions. If problems were found, the engineers and software code writers were readily available to analyze the problem and make repairs. The STA was a complex system that was developed and maintained by gifted staff. I was impressed by their intelligence, work ethic, devotion to NASA and the safety of the astronauts.

I shared a cubicle with a talented engineer, Gerry. He managed multiple programs for the STA team as well as monitoring the oil in the T-38 engines. Debris detection from oil samples enabled Gerry and others to predict and prevent maintenance problems.

Our NASA hosts coordinated a Space Shuttle motion-based simulator session with the STS-7 crew. Mission Commander Bob Crippen, co-pilot Fred Hauck and one of the mission specialists were present. I was fortunate to occupy an empty back seat. The simulator was a tight fit with so many switches, knobs and controls. It boggled the mind to imagine that anyone could know what all the controls did.

The simulator session had blastoff, orbit and recovery profiles. The lift off was exciting with rumbling in our seats as the engines started to produce initial thrust then acceleration as the Shuttle quickly climbed away from the supporting structure. The crew was joyful and meticulous in their management of onboard systems. Simulated failures were introduced to help the crew prepare for possible hazards during an actual mission. Captain Crippen executed emergency procedures from memory. I was amazed at his composure and exactness in the face of failures. The other crewmembers found the emergency procedure in the in-flight guide and checked the correct steps. Invariably, Captain Crippen had completed them all to the letter. His knowledge of the shuttle systems was impressive!

Space Shuttle Simulator cockpit

The STS-8 Shuttle crew planned a trip to Edwards Air Force Base in California for night training. The crew was going to have the first night takeoff and first night landing for the space shuttle program. The STS-8 Commander, Captain Truly, and his copilot, Captain Brandenstein, trained in the STA for night approaches to the Edwards Dry Lakebed. Morri, Nick, Greg and I were invited to go along and we jumped at the opportunity. We flew over to Edwards in the STA. Upon arrival, we went to the Edwards Officers Club for drinks in our flight suits. The room was full of winged aviators and we were excited to be there.

The following night I flew in the STA as Captain Truly made multiple approaches to the dry lakebed. Despite the challenge of the task with easily misinterpreted visual cues and much more difficult environment, Captain Truly executed each approach manually with the precision of being on autopilot. Each approach was flown down to a few hundred feet above the dry lakebed at which point the shuttle simulation training was discontinued and the aircraft flown back up to 40,000 or to the hard runway to conclude the flight.

On another occasion, our group of four ensigns were treated to a tour of the Neutral Buoyancy Lab (NBL)[13]. The NBL is one of the largest swimming pools in the world and enables the astronauts to spend hours practicing space walks and repair procedures underwater in their restrictive space suits. This environment simulates the micro-gravity environment encountered in space. The tour of the facility opened our eyes to another dimension of the astronaut training.

NASA tried to provide each ensign on temporary duty with a flight in the back seat of a T-38. We had all received aviation physiology and T-38 ejection seat training. I received a call from our NASA liaison astronaut, CDR Richards. He mentioned a cross-country flight to Denver and wanted to know if I was interested in going along. I confirmed my great interest.

The T-38 is a sleek looking jet with two turbojet engines with afterburners aligned closely to the center of the aircraft. The wings are quite small in surface area and thickness. The jet looks fast and is fast. Speed and thrust are the keys to its aerodynamics. CDR Richards explained that he needed to conduct training in New Mexico prior to our flight to Denver. We put on our flight gear, stopped into maintenance control and walked to the jet. CDR Richards did the preflight while I got myself strapped into the backseat. He started the engines, taxied us out to the runway and added full power for takeoff.

We were rolling then airborne in no time at all. Flaps were raised and we glided smoothly through the air on our quick climb up to Flight Level (FL) 240[14]. CDR Richards provided me the thrill of flying the aircraft once level at altitude. After a while, he took back control of the jet and we proceeded to land in New Mexico to refuel. The landing in the T-38 was smooth but fast due to the small wings.

After refueling, we re-manned the jet and took off for Denver. The flight to Denver was quick. Denver had afternoon thunderstorms forecast for our arrival. We saw storms cells but were able to dodge them

---

[13] This video provides a solid overview of the NBL at JSC; https://www.youtube.com/watch?v=4mBEaakbzT0

[14] FL240 is achieved by climbing to 24,000 feet Mean Sea Level (MSL) with an altimeter setting of 2992.

and conducted a Visual Flight Rules (VFR) approach to landing. We taxied over to the commercial ramp and parked the jet. I introduced CDR Richards to my family, thanked him for the flight and departed for home. I considered myself extremely fortunate to experience the T-38 aircraft and CDR Richards' aviation skills.

One Sunday morning, STA flight engineer, Marsha, offered a flight in her open cockpit Stearman biplane. Marsha was an alumnus of the CU Boulder aerospace engineering program. I'd never been in an open cockpit aircraft, so the goggles and helmet were novel to wear. Marsha pre-flighted the aircraft, then escorted me into the front seat before taking the rear seat. She started the big radial engine, taxied out to the runway, added full power and had us airborne in a short amount of time. We were flying under VFR conditions and wandered around the countryside in a direction headed for the JSC Headquarters in Clear Lake.

Once back to the airport, Marsha executed a beautiful landing to a full stop and taxied back over to her parking place. She was an excellent pilot and an enjoyable flying companion. In 1984, a year after I left NASA, Marsha was selected to be an astronaut and completed five shuttle missions. While accomplishing many things in space, her long hair going wild in zero G was always worth a smile[15].

Morri and I were discussing Navy bases that conducted primary pilot training. We agreed that getting the base with the easiest flight rules was our goal. We conjectured that the easier training location would enhance our chances of getting good grades and getting the jet of our choice for advanced training. CDR Richards overheard our conversation and provided a pearl of wisdom, "You don't want to attend the easiest training but rather the most difficult. The better your training, the easier it will be to conquer whatever challenges are thrown your way". That comment stayed with me and rang true for everything that followed.

---

[15] Marsha's hair in zero G was always a big hit:
https://en.wikipedia.org/wiki/Marsha_Ivins

# Chapter 3

## Flight Training at NAS Pensacola, FL

*I could no longer be a navy pilot; the dream was gone in an instant with an eye exam.*

Morri and I arrived in Pensacola in January 1984 after driving as a convoy from Clear Lake. Flight school started with a comprehensive medical checkup, general aviation academic training, physical fitness and various survival exercises. The academics included aerodynamics, jet engine design, navigation fundamentals and the human physiology response to flight. The Naval Aerospace Medicine Institute was the first stop for our flight physicals.

I arrived in Pensacola as a navy pilot candidate. The vision requirements for a pilot were 20/20 in each eye, no astigmatism and perfect color vision. My eyes were slightly less than standards eliminating me from the role of pilot. I continued with my class but transitioned to training as a Naval Flight Officer (NFO). NFOs served as Radar Intercept Officers (RIO), Bombardier Navigators (BN), Electronic Counter Measures Officers (ECMO), Electronic Warfare Officers (EWO) and Tactical Coordinators (TAACO) aboard naval aircraft. Once our initial training was completed, the pilots headed off to other Navy bases while student NFOs stayed at Pensacola to train at the squadrons, VT-10 and VT-86.

Water survival exercises were an important part of our initial training. We had to demonstrate treading water with hands elevated for a minute, drown-proofing, breaststroke, backstroke, side stroke and free style swimming. We had to jump off a 15-foot tower into the deep end of the pool and swim 25 meters underwater before surfacing.

The Dilbert Dunker[16] was the device highlighted in the movie, Officer and A Gentleman. It was a source of much anxiety but in

---

[16] Dilbert Dunker historical perspective; https://www.youtube.com/watch?v=aUoLTGQA_jk and with Mayo; https://www.youtube.com/watch?v=_pugcunPOTo

execution wasn't a big deal. The helicopter dunker[17] was a bit more stressful but manageable.

On another day we went to a different pool that had a course set up in the water around the perimeter. Our task was to swim a mile in full flight gear. There were thirty of us that got into the pool at the starting spot. It was crowded and not easy to swim in all that gear. An instructor blew a whistle and we started making our way around the pool. There was a lot of splashing and thrashing as guys made their way around the perimeter. It was a marathon effort and everyone completed the swim.

Once the initial portion of our training was completed, the pilots and NFO candidates parted ways. I didn't see Morri for many years. It was sad to see him go as we had great times together[18]. During our initial training, I met Morri's United States Naval Academy (USNA) classmate, Sam. He and I were both student NFOs and rented an apartment in town.

VT-10, Initial NFO Training

The first day at VT-10 had a surprising twist. Our class size was 30 with both males and females. The instructor running the class was an A-6 Intruder BN with VA-115 experience aboard the USS Midway in Japan. He shared great stories of life in the fleet. To start the class, we introduced ourselves, where we were from and what aircraft we wanted to fly. Most of the class had spoken when it came to my turn to share. I stood up and before speaking the instructor said, "You're going to be an A-6 BN". I replied, "No, I want to fly Tomcats". He replied, "No, you'll be a BN". I shrugged off his comments as he didn't know me but eventually ended up as a BN in his old squadron.

---

[17] Helicopter dunker; https://www.youtube.com/watch?v=92-ThGnmXo4 and https://www.youtube.com/watch?v=POsWGHfHg4M
[18] Morri became a F/A-18 Hornet pilot and led a Navy Hornet squadron in combat before retiring. He shares oral stories from his career here; https://www.readyfortakeoffpodcast.com/podcastepisodes/rft-304-f-18-pilot-ceo-morri-leland

In my class was USNA graduate, Lisa. She lived out at Pensacola Beach with another USNA graduate and student NFO, Barb[19], who was ahead of us in the pipeline. Lisa and I became study partners. We would study together before a test or flight and quiz one another on the course material.

The VT-10 Ground Syllabus taught all the Federal Aviation Administration (FAA) rules that governed flying in the USA along with visual and electronic navigation, radio communications and aircraft specific systems. The flight syllabus had 5 basic flights (B1-B5) in the T-34C, a B-Jet familiarization hop flight in a T-2C Buckeye to experience the jet environment, 5 instrument flights in the T-34C, 9 instrument flights in the T-2C and 5 instrument and one visual navigation flights in the T-39D.

13 June I flew as an observer in a T-39 for 3.8 hours and hoped it would ready me for the B1 flight. The mission was air to air intercepts like those performed by the F-14 Tomcat. I was very interested as Tomcats were my first aircraft choice. Surprisingly, I wasn't impressed with the radar intercepts that day in the T-39. Despite my lack of training, I understood the concept and appreciated the finesse of the aircrew in making the intercept happen. The experience was just rather disappointing. This came into play later at the critical time to select a training pipeline.

Our first graded flight was designed as an intro, yet we were responsible for communicating with the air traffic controllers and using the navigation avionics in the aircraft. All the students in our class started getting nervous as we approached our B1 flight. It evaluated our suitability for military aviation by performing accelerated aerobatics to 4Gs during loops, split Ss, Immelmanns[20], rolls and wingovers. If a student performed well during the initial aerobatics, the pilot executed squirrel-cage maneuvers. This essentially meant he could do whatever he wanted in any attitude, acceleration and airspeed combination provided he did not descend below 10,000 feet Mean Sea Level or exceed 4Gs.

---

[19] Barb had a very successful naval career and wrote 'Flight Lessons' in 2022. https://captainbarbarabell.com

[20] An Immelmann generates a 180° course reversal; https://en.wikipedia.org/wiki/Immelmann_turn

I found the aerobatics during the flight amazing. The pilot started out easy with a 360-degree aileron roll. He then completed a 4G loop that was exciting to experience. Seeing the world from an upside-down perspective was a real eye opener. That led into the Immelmann then a Split S. We did a spin to show what out of control flight might look like. My instructor asked how I was feeling to determine if he needed to continue or finish up. I was feeling fine and encouraged him to continue and do a squirrel cage. That was all he needed to hear. Before I knew it, we were upside down then up then rolling around in every imaginable way while experiencing up to 4 Gs throughout the maneuvers. Once complete, my instructor provided positive feedback on the flight.

The next flights allowed us to practice point to point navigation techniques along with instrument departure and approach procedures. The aircraft used Tactical Air Navigation (TACAN) to provide bearing and range (in nautical miles) to fixed ground or ship-based navigation aids. We practiced Standard Instrument Departures (SID) and Standard Arrival Routes (STAR)[21] procedures using the TACAN. We used TACAN or Approach Surveillance Radar (ASR) for non-precision approaches[22] or the ground-based Precision Approach Radar (PAR) for precision approaches.

The ASR and PAR were called Ground Controlled Approaches (GCA) as an approach controller directed the aircraft's flightpath via radio calls to the crew. The ASR ground controller provided commentary regarding the aircraft's lateral position to the correct approach path to the runway. The aircraft's vertical flight path was managed with a series of altitude step-downs managed by range to the runway threshold. The PAR approach was like the ASR but added a vertical path to landing called a glideslope. The controller told us if we were above or below the ideal flight path as well as left or right of course. The non-precision approach minimums were often above five hundred feet Above Ground Level (AGL) while the PAR approach allowed a descent to 200 feet AGL

[21] An overview of SIDs and STARs can be found here; https://aerocorner.com/blog/sids-and-stars/
[22] An overview of instrument flight procedures can be found here; https://www.faa.gov/regulations_policies/handbooks_manuals/aviation/instrument_procedures_handbook

provided the pilot flew an accurate approach. The lower altitude was essential to see the runway environment in poor weather like a low ceiling, limited visibility, fog, snow or heavy rain. If the runway environment couldn't be acquired in time to complete a safe landing, the pilot was required to add full power and climb away from the runway on a missed approach. Another approach could be attempted, or the aircraft could divert to an airport with better weather.

At the end of VT-10, it was time to select which aircraft pipeline to pursue. I initially wanted to be an F-14 RIO but had a change of heart from my start at VT-10. I'd flown a radar intercept flight in the T-39 and was not impressed by the role. I'd also loved flying low and fast to the ground in the T-39D. I requested the Tactical Navigator training for my transfer to VT-86 and fortunately was accepted.

VT-86 Advanced NFO training

*'First sight, win the fight. Lose sight, lose the fight'*
*Rules for Air Combat.*

T-2C Buckeye, TA-4J Skyhawk and T-39D Sabreliner aircraft were flown at VT-86. The T-2C was a versatile, twin-engine jet aircraft with a large, comfortable cockpit and predictable flying qualities. The TA-4J was the trainer version of the single engine, carrier-based attack plane that saw combat service in Vietnam. It was also flown by the Navy Flight Demonstration Team, the Blue Angels[23]. The TA-4J cockpit was quite narrow requiring me to rotate my torso as the pilot lowered the canopy to ensure that I didn't get my flight suit or body parts pinched. The T-39D was a small, older business jet with dated avionics. The T-39D PIC sat in the left seat and students sat in the front right seat or in the back during radar navigation or intercept practice.

While in the jets at VT-86, we practiced our SIDS and various instrument approaches at higher speeds. We also trained to fly low level navigation flights in the T-39 using visual cues out the front windscreen or using the ground mapping radar. The flight paths were standard military training routes that had vertical and horizontal boundaries to

---

[23]https://www.history.navy.mil/content/history/museums/nnam/explore/collections/aircraft/a/a-4-skyhawk.html The TA-4J was also known as the 'Scooter'.

permit high-speed, low-level flight. Normally, aircraft need to be under 250 KIAS below 10,000 feet MSL in the continental USA. These routes often allowed speeds up to 540 KIAS at altitudes down to 200 feet above the ground. Our flights in the T-39 were normally flown at 300 KIAS and 500 feet AGL. We used 1:250,000 topographical maps and labeled them with headings, distances and time between waypoints. We produced navigation cards with headings that we updated with wind corrections after receiving our preflight weather brief. We learned to rely on our wind corrected preflight headings and leg times to navigate if visual landmarks were unavailable.

The radar navigation flights were less stressful as we were further away from the ground but more difficult as we had to use the radar to interpret ground returns to maintain navigational situational awareness. The radar had rather average resolution but could enabled us to identify a bridge over a river in a town. Preflight study of our map along with predictions of what we would see helped considerably. The skills of time, distance and heading learned on visual low levels served us well.

My last three flights at VT-86 were in the TA-4J with a two-plane cross country to NAS Key West and MacDill AFB, south of Tampa, FL. Marco was the student NFO in the other aircraft. He was headed off to the F-14 community when he finished VT-86[24]. We got along well as classmates and looked forward to being away from Pensacola. The focus of the three flights was to train in one vs one dogfighting tactics. We had been taught the classroom theory of offensive and defensive Air Combat Maneuvering (ACM)[25] and these flights were the practical lab. We had been taught standard phraseology for various maneuvers and were tasked with observing the inflight scenario and making the correct calls to our pilot. The instructors acted as NFO-commanded autopilots executing the maneuver requested, safety permitting. This proved to more challenging than expected. It was difficult to speak clearly and intelligently while

---

[24] Marco made the big screen in the movie TOPGUN and enjoyed a successful career in the Navy; https://twitter.com/StacyPearsall/status/1271178400787566594/photo/1
[25] OPNAVINST 3710.7 defines ACM as 'Aggressive three-dimensional maneuvering between two or more aircraft simulating offensive or defensive aerial combat where the potential for a role reversal exists; or defensive maneuvers or other combat avoidance maneuvers by one or more aircraft. Numerous rules to safely manage ACM are listed in 3710.7.

pulling and sustaining up to 6.5Gs. The situation was dynamic so a delay of a second in execution meant the difference between gaining the offensive or being on the defensive. An inappropriate call easily resulted in giving our opponent a simulated guns shot on our aircraft.

Thursday 20 December The takeoff from Pensacola was exciting. As a single engine aircraft without afterburners and a small wing, the TA-4J needed a fair amount of airspeed to get airborne. Once the gear and flaps were raised after takeoff, the jet's maneuverability started to shine. We flew as a two-plane section down to the Warning areas off Key West. Once there, we performed G awareness maneuvers to ensure that our G suits were working and that we were feeling ready to fight the aircraft. We split up our flight with headings 180° away from one another for several minutes. We coordinated our turns back into each other over the radio as we could no longer visually see each other, and the A-4 did not have a radar. The motto for air combat was, 'First sight, win the fight. Lose sight, lose the fight'.

I strained to visually acquire Marco's aircraft so I could change our course and take away any angle advantage that he might have. ACM has a lot to do with angles. If you give up angles in your position, the enemy uses them to gain a tail aspect and shoot you down. You tried to gain angles on your opponent to gain that tail aspect on them for a shot[26]. This all seemed simple until trying to do it with a closing velocity of 1000 nautical miles per hour.

Once we spotted the other aircraft, we turned to put our nose on them to intercept yet maintain our 3710.7 mandated, 500 feet safety bubble around our aircraft. Five hundred feet of aircraft separation isn't much when traveling at those closure speeds. The aircraft were generally flying in one G until our aircraft passed one another. As we passed, the pilots snapped on 6.5G sustained turns to gain those elusive angles over the opposition. This generated another intercept passing of the aircraft where the 500-foot bubble once again needed to be observed. Another maximum G turn was used to gain more angles on the opponent. This occurred again and again usually resulting in a three-dimensional accelerated maneuver called a rolling scissors where neither pilot had an

---

[26] ACM basics are presented well here;
https://www.youtube.com/watch?v=At3qlnd_Ugo

advantage, but both continued to seek advantage of the other due to mistakes or fatigue[27]. There were no clear winners in our combat setups, but we all received a great workout under the G loads. We proceeded to NAS Key West, arranged for the jets to be refueled, debriefed the flight and checked into the BOQ.

Friday 21 December We had breakfast and checked out of the BOQ. We briefed the second ACM flight and headed out to the jets for takeoff. Our plan was to return to the Warning areas for more ACM then land at MacDill AFB, near Tampa. The ACM was like the day prior, but Marco and I were both feeling more comfortable with what to expect, what to do and say. There were no clear winners again and the engagements once again turned into rolling scissors. Once done with the ACM, we joined up and headed over to MacDill. We landed, arranged for fuel, debriefed and headed over to the BOQ once again to secure rooms.

Saturday 22 December We repeated the morning ritual of breakfast, BOQ checkout and briefing for the flight back to Pensacola. It was a straight shot across the Gulf of Mexico to get home with a stop in the Warning areas enroute to complete more ACM. Marco and I swapped instructor pilots for the last flight. I had a salty, fleet experienced F-14 pilot, callsign Eagle. He was great to fly with and I felt an advantage when it came time for the ACM engagements. As this was the last TA-4J flight for Marco and I, the instructors felt more freedom to really fight their aircraft to the best of their ability without the perpetual student dialogue of canned phraseology and maneuvers. Marco and I were able to experience a pair of tactical naval aviators unleashed on one another. It was a sight to behold. The A-4 was such a sporty, nimble aircraft to fly and fight. It was an honor to complete my training at VT-86 with a fleet experienced fighter pilot showing me the skills of his trade. We finished the ACM, landed at NAS Pensacola and debriefed the flight completing our training for Naval Wings of Gold!

The winging ceremony was at the Naval Aviation Museum on NAS Pensacola. There were 15 of us 'winged' on the day. I was excited to be selected for Intruder BN training at NAS Whidbey Island, WA. As I headed off for Whidbey Island, Sam stayed behind to complete

---

[27] Rolling Scissors are discussed here;
https://www.youtube.com/watch?v=2qAPlMbTulM

Electronic Warfare (EW) school. He was selected to become an ECMO aboard EA-6B Prowler aircraft.

Before Whidbey, I stopped at Survival Evasion Resistance Escape (SERE) school in San Diego. SERE had a few days of classroom instruction followed by a four-day, three-night survival exercise/prisoner of war camp practice. The classroom instruction provided the history of the treatment of prisoners of war and strategies to use to manage the experience. Our instructors taught us how to work together as a team if placed in a prisoner of war scenario. The experience was a valuable lesson in survival with various physical, emotional and psychological challenges.

# Chapter 4

## Intruder Training at NAS Whidbey Island, Washington[28]

*Axel executed a bombing maneuver called a popup. We raced toward the target but with 3 nmi to go, he abruptly cut 45 degrees to the right, climbed to 2000 feet, rolled left to upside down, pulled the stick to place the aircraft's nose on the target, rolled upright into a 15° dive, designated the target using his gunsight then released the bomb.*

The last leg of the western journey was along I-5, past Seattle to state road 20 leading to Whidbey Island. This was the location of the west coast Fleet Replacement Squadron (FRS) for A-6E aircraft, the Golden Intruders of VA-128. Upon arrival at Whidbey Island, a large billboard leading to the Naval Air Station had pictures of an A-6E Intruder[29] and an EA-6B Prowler with the words "Pardon our noise. It is the sound of freedom". I heard those words many more times in my naval career.

I reported aboard NAS Whidbey and checked into the BOQ with a room on the second deck. It had a beautiful view above the Officers Club toward the airfield and water. Washington State is covered with lush green trees as it rains throughout the year. The room was simple but served me well for the next few months of study.

The Intruder FRS taught Fleet Replacement Pilots (FRP) and Fleet Replacement Bombardier Navigators (FRBN) how to fly and fight their aircraft. FRPs and FRBNs were trained simultaneously on the same topic to economize resources. For example, a group of FRPs and FRBNs were taken to NAS El Centro for a Visual Weapons detachment for ten days and to the aircraft carrier to qualify landing the Intruder aboard the ship.

While at Whidbey Island, FRPs gained experience flying the Intruder in close formation with other aircraft. FRBNs learned how to navigate and attack targets using the radar. Once that skill was developed

---

[28] Pictures of A-6E Intruder and TC-4C aircraft.
https://www.seaforces.org/usnair/VA/Attack-Squadron-128.htm
[29] Intruder summary; https://www.youtube.com/watch?v=s0N6SsKivxU ;
https://en.wikipedia.org/wiki/Grumman_A-6_Intruder

and refined to hit discrete buildings in a dense urban environment, the FRBNs were trained to use the straw-like, fine targeting of the onboard Forward-Looking Infrared Detector (FLIR) with its onboard Light Amplification by the Stimulated Emission of Radiation (LASER) rangefinder. The FLIR and LASER were part of the Target Recognition and Attack Multi-Sensor (TRAM) located in a ball under the nose of the Intruder called the Detection and Ranging Set (DRS) turret. The DRS also contained a LASER Receiver and Designator. The Receiver allowed the BN to see a LASER spot designating a target from another aircraft or ground-based soldiers. The Designator allowed the BN to designate a target for LASER guided weapons. The FLIR and LASER ranging made the Intruder a bullseye bomber.

FRPs and FRBNs flew day/night low levels with instructors and learned how to attack targets in all weather conditions while at low altitude to penetrate enemy defenses. This training transitioned FRBNs into the 'Dark Wizards' described well by Intruder pilot, Francesco 'Paco' Chierici[30]. The BN radar controls are below.

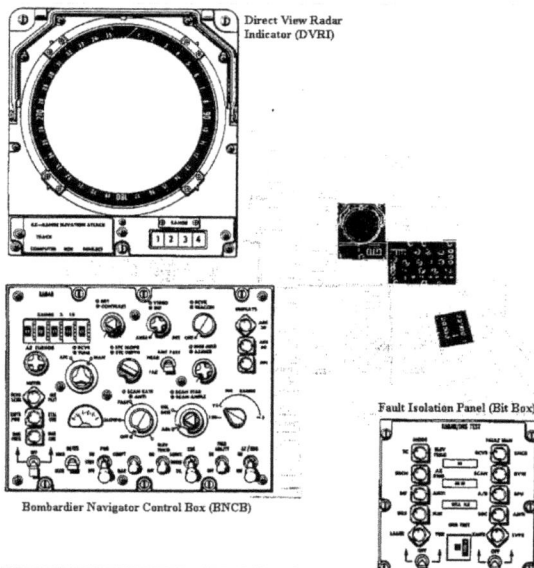

Direct View Radar Indicator (DVRI)

Fault Isolation Panel (Bit Box)

Bombardier Navigator Control Box (BNCB)

[30] Paco shares his insights; https://www.thedrive.com/the-war-zone/27604/confessions-of-an-a-6-intruder-pilot

The A-6 Intruder cockpit with the BN slew stick and control panels for the armament, computer, radar and FLIR and in the foreground.

Many of our class FRBNs and FRPs met on our first day of class. They all had a good sense of humor and were happy to be there. That was important for the challenges ahead. We had various officers brief us about how flight operations worked in the squadron. The second in command, Executive Officer (XO), CDR Steve 'Axel' Hazelrigg, welcomed us. He was a John Wayne type character with a quick and bright smile, deep voice and great confidence as a navy test pilot. Axel was a large, strong man with a firm handshake and no-nonsense manner that was endearing. He opened with the motto, "Work hard, play hard"[31]. As new crew, we were told that the FRS was different to the training command where we earned our wings. The FRS was designed to teach us how to fly the Intruder most effectively, but they weren't interested in losing any aircrew to attrition. It cost too much to get us to the FRS and they didn't want to lose the investment. That helped us relax slightly but we still had a lot of work to do.

---

[31] I heard that phrase many times during my career from many great leaders.

They handed us the three-inch-thick Naval Aviation Training and Operating Procedures Standardization (NATOPS) manual and a one-inch-thick Pocket Checklist (PCL) for the Intruder and told us to read them cover to cover and memorize the numerous Bold Face Emergency Procedures (EP), prohibited maneuvers and operating limits for the aircraft. The Bold Face procedures required immediate action to protect the aircraft and crew from significant danger. We were scheduled to sit a closed book test after three days but had no duties prior to that. We needed to memorize the EPs verbatim and couldn't miss any steps on the test. We all left the squadron and began studying.

NATOPS manuals were nicknamed the 'Giant Blue Pill' as they put many aviators to sleep with their technical writing style and in-depth detail about operating systems. Absorbing the whole manual in three days was overwhelming. We had to study EPs and limits with the aircraft while in the training command but we weren't NATOPS qualified in those aircraft. This was our first NATOPS qualification. We needed to transform into Intruder subject matter experts. I made up 3 x 5 cards with EPs and aircraft limits on them. I started doing this at the beginning of flight school and found it to be an excellent way to study. Learning started with making up the cards, but my knowledge was solidified by going over them many times. It took time to make up the cards, but they fostered ownership of the information in my mind. I reviewed the EPs and limit cards before every flight to refresh the information in my mind. We were expected to spend a considerable amount of time studying to learn all about the aircraft, its propulsion, hydraulic, fire suppression, electrical, flight control, fuel, avionics, computer, emergency egress and life support systems.

The test after three days of study proceeded well for the class. We were all pleased to have survived that initial challenge.

The weather around Whidbey Island was generally rainy 10 months of the year with grey, overcast clouds. During the winter, the clouds were often quite low and mixed with fog. This provided excellent training for all weather attack aircrews but it was cold and wet! It was not uncommon to enter clouds within 30 seconds after takeoff and depending on the mission, you might not see the ground again until just before landing. In the months of July and August, the skies were blue, the weather warm and it was glorious. The scenery was stunning including the ocean surrounding the island but the water temperature is 50° Fahrenheit year-round. We weren't going swimming in it despite its visual temptation. If we ejected over water, we needed to quickly get into our personal survival raft while our hands functioned. It we didn't, we could die from hypothermia before the rescue helicopter could be on scene to pick us up. This fact was refreshed every day in our preflight briefings.

Bing was a great friend from flight school in Pensacola and was headed up to NAS Whidbey Island to fly Intruders as well. While in Pensacola, we spoke positively about living together while at Whidbey. It was agreed that I'd live in the BOQ until his arrival in April when we would look for a place to rent. Bing found another friend, Bill, who was also headed to Whidbey, so we found a three-bedroom, rental house.

One of the first tasks that the FRBNs needed to master was making low level navigation charts. There was a community standard that was taught explicitly, and we were expected to follow it. As the Intruder aircraft didn't have a moving map display, the BNs kept track of the aircraft's position on a paper map. A kneeboard card with times, distances, headings and fuel data was developed to supplement the flipchart. This process took many hours to manufacture for a thirty-minute low level. Every low level we flew needed one of these charts and cards. Countless FRBN hours were spent making charts while the FRPs learned the skills of how to fly the Intruder.

Before flying in the aircraft, we needed to demonstrate in the simulator that we were familiar with the aircraft procedures, checklists and EPs. I was crewed with one of my pilot classmates, KO, for our first Weapon Systems Trainer (WST). It was the first of fourteen WST events for me while a student at VA-128. KO was very intelligent and had done a great job memorizing the EPs including switch/control manipulations. When the WST console instructor gave us an emergency, KO was right on top of it nailing the procedures. That made my job easier as I verified his actions and pulled out the PCL for confirmation of the accomplished and additional steps to be followed. After two hours of EP torture, we had successfully completed the checkout and were cleared to fly the aircraft.

We went over to the schedules office and informed them of our simulator completion. I was on the flight schedule for the next day. KO flew his first few flights with an instructor pilot in the right seat instead of a BN. This was a precaution to make sure a FRP comfortably transitioned to the aircraft before flying for the first time with a BN.

Tuesday 2 April My first flight in the Intruder occurred after a lot of study, in-person training in class, the WST supervised by an instructor, exams and assessments. It seemed like the day would never arrive. The flight instructor helped me feel comfortable in our preflight brief. We examined the jet and got into our ejection seats. I immediately got to work setting up each control to the proper prestart position. We added external power to the jet and I started entering the latitude and longitude of our present position to commence the alignment of the Inertial Measurement Unit (IMU).

The IMU was a piece of avionics that contained gyros and very sensitive accelerometers operating in three perpendicular axes. The accelerometers fed their data to the Inertial Navigation System (INS). The INS maintained the aircraft's position without any external references. We entered target coordinates, the computer led us into the target area and we used manual targeting with a gunsight aimpoint, radar cursors and Forward Looking Infrared (FLIR) reticle to designate the target aimpoint.

The IMU alignment took 6.1 minutes and we were ready to taxi. We taxied out to the radar warmup area where we used a radar reflector in the distance to verify the accuracy of the weapon system. BNs could place their radar cursors on the reflector and the pilot measured the distance from his gunsight reticle to the target center in milliradians – abbreviated mils[32]. The aircraft's bombs impacted where the pilot's reticle was aiming. The BN could slew his FLIR reticle on the reflector and the pilot would repeat the exercise. A perfect jet would have the radar cursor, the FLIR reticle and the pilot's gunsight reticle all dead center on the radar reflector. This rarely happened but typical errors of 3 mils didn't generate significant bombing errors. If an aircraft had significant

---

[32] 1 milliradian = 6 feet at 6000 feet. https://gununiversity.com/understanding-mils/

system alignment errors, that aircraft would be grounded for bombing missions but could fly other training flights.

Once we were cleared for takeoff, the pilot pulled out onto the runway centerline and applied the brakes while he added full power. We looked at the engine instruments to ensure they were within operational limits while the pilot cycled the flight controls to ensure free movement and no hydraulic anomalies. The pilot checked with me to verify I was ready to go, then released the brakes. Our acceleration was slower than in the training command aircraft with a longer takeoff roll. Once airborne, the pilot raised the landing gear as we commenced our climb away from the runway. Tower gave us a frequency change as we cleared their airspace and we contacted the departure controller.

As we continued the climb, I turned on the ground mapping radar and put my face up to the boot to examine the image clarity. The radar boot was a light-screening, plastic shield with a foam ring at the top. The BN rested his face on the ring as he looked at the radar and FLIR screens. The boot kept daylight off the radar screen enabling the BN to see the displays better. The boot also kept the radar and FLIR light from escaping at night to protect the pilot's eyes from night blindness. As I peered into the boot and saw mountains on the radar screen, I was immediately impressed. The radar resolution was much better than the T-39 air to ground radar. The picture looked three-dimensional and I looked forward to using it.

VA-128, VA-42 (the Navy's east coast FRS) and the USMC FRS, VMAT (AW)-202, used modified Grumman G1 aircraft called the TC-4C Academe[33] to train the Intruder FRBNs. The TC-4C was a twin turbine powered aircraft with an Intruder nose containing the search radar and noticeable DRS turret below. Inside the rear of the aircraft was an Intruder cockpit that allowed a FRBN to practice radar navigation and bombing while the supervising instructor could induce weapon system failures. The instructor watched the FRBN as he analyzed a malfunction, correctly downgraded his weapon delivery methods and completed the mission. FRBNs flew eleven times in the TC-4C during their training at the FRS. The VA-128 flights started at Whidbey and proceeded east at

---

[33] Grumman TC-4C training aircraft.
https://en.wikipedia.org/wiki/Grumman_Gulfstream_I

high altitude over to Spokane, WA. The intelligence department generated high-resolution pictures with accurate latitude/longitude/altitude data for various landmarks in Spokane. FRBNs were given this information then required to generate radar predictions of what the targets would look like from various distances and run-in headings.

Simulated attacks were conducted against the targets using the attack mode of the mission computer. A Radar Bomb Scoring (RBS)[34] system allowed a ground controller to estimate a simulated bomb impact point. A bomb tone was broadcast on the UHF radio from the TC-4C aircraft identifying when a simulated bomb would release. A ground tracking station could then process the aircraft's position, altitude and speed with the release time and predict a weapon impact location. The runs were made at 240KTAS due to the speed limitations of the TC-4C. Competitions were held each month to see who could get the best bomb scores at Spokane. When first arriving at Spokane, FRBNs learned to identify various targets starting with 'Boomer Bridge'. It was called Boomer as you couldn't misidentify it with its large radar return, position in the city and orientation to the river below.

Our TC-4C aircraft flew back and forth over Spokane attacking target after target with 30nmi runs. We flew above 10,000 feet MSL in day, night and poor weather conditions. At no point was there ever a risk of any ordnance coming off the aircraft as it didn't carry any. Few people in Spokane were aware how we were using the city layout to expertly train the FRBNs in radar bombing. Despite this, Spokane provided an invaluable service to the U.S. Navy in training its Intruder crews. Prior to completing the FRS, each FRBN was given the opportunity to fly the faster Intruder against the Spokane targets provided that the aircraft was not carrying any ordnance. Well-trained FRBNs looked forward to this Spokane flight as it allowed them to use their radar and FLIR at speeds more representative of combat. As a BN transferred his target designation from the radar to the FLIR, the pilot's steering symbology moved to arrive at the correct weapon release point. The mission computer took in airspeed, altitude, G loading, dive or climb angle and wind corrections to provide pilot steering commands to reach the bomb release point. Later

---

[34] RBS details; https://en.wikipedia.org/wiki/Radar_Bomb_Scoring

in the fleet, we created practice targets in urban areas to keep current in finding discrete targets among dense radar returns.

Friday 3 May FRPs and FRBNs trained in visual low level flight routes at 200 feet AGL and 360 Knots of Ground Speed (KGS) (equates to 6 nmi per minute). We increased our speed to 420-480 KGS (7-8 nautical miles per minute) as we attacked a target. My first low level flight in the A-6E was with Axel. He was a man of few words that enjoyed sharing his great love of flying.

We took off from NAS Whidbey and checked in with Seattle Center to cross the Cascade mountains to the east. They were hidden by clouds requiring us to stay at altitude until the weather cleared. Once in Visual Meteorological Conditions (VMC) conditions, we cancelled our Instrument Flight Rules (IFR) clearance and proceeded visually with a descent to 200 feet AGL. Axel's gifts as a pilot came alive at low altitude. He gracefully carved his way through the sky with hills, mountains and trees on either side of him. At times, we would follow a narrow river valley while pulling up to 5 Gs to stay low and inside the ridgelines.

The low level took about 25 minutes to arrive at our practice bombing range at Boardman in northern Oregon. As the flight was designed as a demonstration for me, Axel checked in the bombing range control tower and proceeded to make passes at the large bullseye on the ground. He started low with a level laydown followed by a 6.5G pull away from the target. The bombs dropped were blue MK-76 practice bombs that weighed 25lbs (12.5kgs). They had a smoke charge for day drops and a flare charge for night drops that allowed the tower to mark the accuracy of the impact. The 6.5G pull away from the target was the maneuver that Intruder pilots executed as habit to get away from any fragmentation from the explosion of an actual bomb.

Axel executed a bombing maneuver called a popup. We raced toward the target but with 3 nmi to go, he abruptly cut 45 degrees to the right, climbed to 2000 feet, rolled left to upside down, pulled the stick to place the aircraft's nose on the target, rolled upright into a 15° dive, designated the target using his gunsight then released the bomb. Axel commanded another 6.5G pull and back down to 200 feet to egress away from the target.

He requested the high pattern and we climbed up to 10000 feet AGL in an orbit above the bullseye. Once approaching the weapon delivery approach course line, Axel transitioned from his 30° angle of bank in orbit to nearly upside down as he pulled the nose of the aircraft toward the ground. He had planned his distance abeam the target precisely so as we rolled upright, we were in a 40° dive toward the target with our airspeed rapidly increasing. This bombing maneuver was not designed for those faint of heart as we accelerated to 500KTAS in the dive. Axel released the bomb then snapped on another 6.5G pullup to get away from the rapidly approaching ground. I was amazed how he quietly and confidently maneuvered the aircraft into position, called the tower, released the bomb, made a radio call to confirm release then started the whole process over again. The bombing pattern was second nature for Axel but a real eye opener for me.

Once finished at Boardman, the weather was good enough to fly a portion of the VR-1355[35] visual low level that headed back towards Whidbey. The 1355 was nicknamed, 'The Million Dollar Ride' due to its path through the most beautiful terrain in Washington's Cascade mountains. The route flew by peaks that rose to over 10,000 feet MSL. We headed northwest to enter the route at point B and Axel set our radar altimeter to 190 feet. We rounded the corner at point B at 3000 feet MSL and headed toward point C, Rimrock Lake.

Axel seemed to find every nook and cranny to lower the MSL altitude of our aircraft. As we approached Rimrock, a nice valley presented itself slightly west of the route. We could be up to 4nmi from the route centerline so this worked well. As we approached the lake, Axel dove down to the left to fly up the lake. At the west end of the lake, we made a right 90° turn coming to a northerly heading toward point D, Keechelus Lake along I-90. Flying this low level with Axel was like watching a beautiful ballet. His maneuvers in the aircraft were graceful and deliberate. As we approached point D, we deviated to the east to

[35] https://www.youtube.com/watch?v=6ELJkITPr0A VR-1355 from an EA-6B front right seat in March 2010. https://www.youtube.com/watch?v=N4-VHMkHEUQ&t=17s VR-1355 from F/A-18G back seat..

avoid 500 feet AGL logging cables and I-90 along the lake. Apparently, motorists along I-90 get rather distracted while watching Intruders racing parallel to them down the lake at 200 feet. We stepped up our altitude as we overflew I-90 at a perpendicular angle then settled back down onto Kachess Lake and headed north into the Alpine Lakes Wilderness Area.

The high mountain lakes were a stunning deep blue color. It would be a great challenge to reach them on foot as there were no roads in sight. As we climbed up to meet the ridge lines at 8000 feet MSL, Axel added power early to ensure we kept our airspeed and maneuverability. The aircraft wasn't as snappy at higher altitude as it was back at the start of the low level. As we headed north, clouds were obscuring the route and we climbed above them, contacted Seattle Center and returned to Whidbey Island. That had been an amazing ride in the Intruder. I felt very lucky to be in the community and training with such a great instructor.

Saturday 8 June NAS Whidbey Island had an Aero Club that enabled military personnel an opportunity to learn to fly and gain their private pilot's license. My goal was to become a better BN by understanding the responsibilities and flying techniques of pilots. With 9.6 hours of dual instruction time, my instructor got out of the C-150 aircraft let me complete 3 solo landings. That was a big thrill. I completed solos at Mount Vernon and Oak Harbor airports. Oak Harbor airport was a challenge as the runway was only 25 feet (7.6m) wide! There wasn't much room for error. It made me a better pilot and gave me more appreciation for landings on aircraft carriers.

Night or IFR, low-level flight in mountainous terrain was a unique, trademark capability of the A-6E Intruder. It was preferable to fly at night in combat to defeat many visually acquired threats. It was preferable to fly fast at low altitude to stay below enemy radar horizons, to aide in tactical surprise and minimize the threat of enemy fighters and surface to air missiles. We flew three low level routes at night while students at VA-128: the IR-342, the IR344 and the IR-346. These instrument routes could be flown during the day, night and in any weather. We flew these routes at 360 or 420 KTAS.

Generally, the only American aircraft that would fly these routes in instrument conditions or at night were the Intruder, the Air Force F-

111 and the F-15E Strike Eagles. Many crews had died on night low levels over the years, so the procedures and scheduling were written in blood. Prior to flying a night low level, students needed to fly the identical route in a simulator, then during the daytime followed by the night flight within seven days with the same crewmember as flown during the day. If the night low level didn't happen within 7 days, another day flight of that route was required. The IR low levels normally lasted thirty or forty minutes with an attack at the Boardman bombing range followed by the 1355 low-level home during the day or a high altitude return at night. The attack at Boardman included dropping one MK-76 practice bomb.

The Intruder had a software mode in the mission computer called Search Radar Terrain Clearance (SRTC)[36] that correlated data obtained from the search radar onto the pilot's Attitude Display Indicator (ADI) in a 3D-like depiction of the terrain ahead of the aircraft out to 3nmi. The ADI allowed the crew to comfortably fly up to 480KTAS, 1000 feet AGL through mountainous terrain at night, in fog, heavy rain or snow. The pilot could select buttons on the display that would enable the crew to fly as low as 300 feet AGL in benign terrain but it took more crew coordination and even more strict adherence to procedures. Prior to flight, the BN produced navigation charts that were specific to each low level with timings, distances, and headings for each leg of the flight. Prior to preflight brief, the FRBNs studied the terrain and practiced the calls that they would make to the pilot as they were flying the route. When traveling at 7 or 8 nmi every minute, the crew needed to think ahead of the aircraft to stay safely away from the ground.

There was a lot of talking in the Intruder cockpit on a night low level. The communication was standardized as were the procedures for the crew to safely navigate through mountains. The BN had a wider and deeper field of view than the pilot's display so he could add to the pilot's situational awareness. The pilot and BN would confer on climbs that were required to clear upcoming terrain. Once a mountain had been cleared, the BN would see terrain in the distance, call it to the pilot and he stopped his climb. They would both wait for the radar altimeter to register the ridge crossing with a low reading followed by increasing

---

[36] Intruder crews referred to SRTC low level flying as Terrain Clearance (TC).

values before the pilot commenced a descent. The BN ensured there was terrain clearance in turns as he could see 55° from the aircraft centerline while the pilot could only see 15°. Without a BN's help, a pilot could turn into a mountain ridge without seeing it on his ADI until it was too late to ensure clearance.

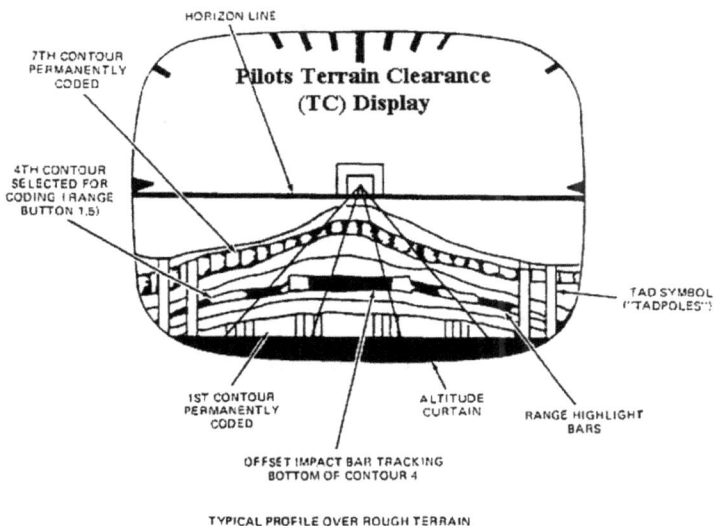

HORIZON LINE

7TH CONTOUR PERMANENTLY CODED

**Pilots Terrain Clearance (TC) Display**

4TH CONTOUR SELECTED FOR CODING (RANGE BUTTON 1.5)

TAD SYMBOL ("TADPOLES")

1ST CONTOUR PERMANENTLY CODED

ALTITUDE CURTAIN

RANGE HIGHLIGHT BARS

OFFSET IMPACT BAR TRACKING BOTTOM OF CONTOUR 4

TYPICAL PROFILE OVER ROUGH TERRAIN

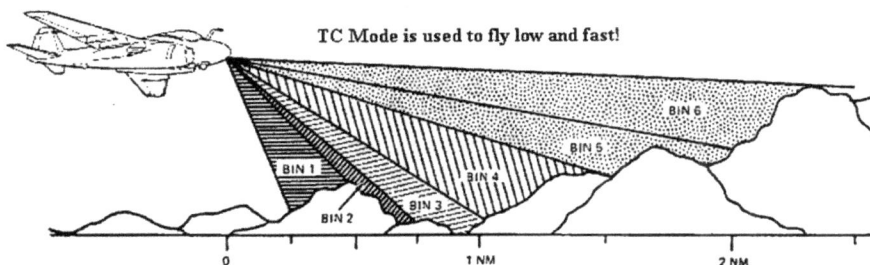

TC Mode is used to fly low and fast!

BIN 6

BIN 5

BIN 1

BIN 4

BIN 2

BIN 3

0          1 NM          2 NM

Wednesday 12 / Monday 17 June I successfully completed the IR-342 day / night low level. The instructor pilot was the squadron flight surgeon, callsign Doc. He was a very rare breed of Navy Medical doctor who had his Navy pilot wings, was NATOPS qualified in the A-6E Intruder and was an instructor pilot. We proceeded at high altitude over the Cascades to the low-level entry point south of the Boardman bombing range. Our standard low-level discussion started as we conversed about our initial descent, radar altimeter settings, the terrain ahead, climbs, ridge passing, descents, time to turn-point, outbound headings, terrain in

43

the turn and a myriad of other items. The day flight proceeded well and we repeated the route at night five days later. It was amazing to be flying at low altitude at night and very rewarding.

Many people ask what it is like landing aboard an aircraft carrier. My response is usually, "Exciting!" It is a combination of exciting and anxiety inducing[37].

Before discussing landing aboard an aircraft carrier, some background information is required[38]. Approaches to landing at the carrier came in three primary types: Case I, Case II and Case III. The Case II and III approaches were used when the weather demanded an instrument approach.

The Case I pattern was used during the day when the atmospheric conditions were greater than VFR weather minimums[39]. Aircraft entered the pattern from overhead the carrier at 800 feet MSL and initiated an accelerated left turn (the break) to enter a downwind that paralleled the ship's course and was slightly over 1nmi from the carrier at the closest point.

The aircraft slowed down in the break and transitioned to the landing configuration with gear and flaps down on the downwind while descending to 600 feet MSL. The downwind altitude of 600 feet was lower than most fixed-based airport flight patterns. The turn to final was initiated when the aircraft was abeam the stern of the aircraft carrier. Approximately 25° left Angle of Bank (AOB) and 300 Feet Per Minute (FPM) rate of descent was used on the continuous turn to Final. The altitude after 90° of turn was 450-500 feet MSL and was 375 feet MSL rolling onto Final (in the groove). Meticulously abiding by these numbers helped pilots get a good start on their approach eliminating the need to make dramatic corrections when close to landing. Once in the groove, the pilot maintained the optimum glideslope using his engine throttles.

---

[37] The aircraft carrier is the most exciting place to work;
https://www.youtube.com/watch?v=q4OnvoXLMxQ
[38] Carrier Operations naval instruction; https://info.publicintelligence.net/CV-NATOPS-JUL09.pdf
[39] Case I;
https://upload.wikimedia.org/wikipedia/commons/b/b4/US_Navy_Day_case_1_landing_pattern.jpg

The carrier's Fresnel Lens system provided a visual indication to the crew of the aircraft's position in the sky relative to the desired 3.5° glideslope to landing. A yellow ball aligned with green datum lights indicated that the aircraft was on the desired glideslope. If the ball was seen above the datum lights, the aircraft was above the glideslope. If the ball was below the datum lights, the aircraft was below the desired glideslope. If the ball was well below the datum lights and turned red in color, the aircraft was dangerously below the desired glideslope and could crash. A red ball demanded an immediate application of more power to climb and avoid an accident.

The pilot used lateral stick movements to maintain alignment with the landing area centerline and used forward and aft stick movements to control the aircraft's pitch and approach airspeed. The time that the aircraft was wings level in the groove was usually between 15-18 seconds. During this short period, carrier pilots make a considerable number of small corrections with their stick and throttles[40] to successfully complete the approach and land with such a large force that it would shock even the hardiest airline passenger.

Case II approaches started as Case III then transition to the Case I pattern when the crew was in visual contact with the carrier inside of 10nmi. Night and poor weather approaches were Case III[41] and they always produced anxiety in anticipation and execution. Prior to commencing a Case III approach, crews were sent to holding patterns that were stair-stepped in altitude and range from the carrier[42]. Each crew was provided with a holding radial from the carrier, a specific altitude to hold, the current altimeter setting, the expected final bearing and the time to commence their approach (whole minutes were used). The holding pattern was established with a fix on the holding radial at a distance that equaled the assigned altitude plus 15nmi (referred to as Angels + 15). From this fix, the crew flew a counterclockwise racetrack holding pattern

---

[40] A carrier pilot's landing control movements; https://www.youtube.com/watch?v=NpXkxG7vnAI&t=22s
[41] Case III approach; https://i.redd.it/pt4eq0grwzb11.png
[42] Called the Marshal stack, the Intruders held at the highest altitude and furthest from the carrier. On clear nights from that perch, we saw the other airwing aircraft synchronized below us.

with 6nmi legs until their commencement time. The pilot modified the holding pattern as needed to start the approach exactly on time from the holding fix[43]. This was often a challenge as the aircraft arrival at the fix was to be within 5 seconds of the assigned approach commencement time. Approach control needed to be notified if the approach commencement deviated by more than ten seconds from the assigned time.

Upon commencing a Case III approach, crews flew inbound to the carrier at 250 KIAS with a 4000 FPM Rate Of Descent (ROD) until 5000 feet MSL (called Platform), when the ROD was slowed to 2000 FPM until reaching the final approach altitude of 1200 feet MSL. This normally occurred prior to 10nmi. Aircraft transitioned to the landing configuration with gear and flaps down normally at 10nmi but always before 8nmi. Once in the landing configuration, crews slowed to their landing approach speed, normally 132KIAS for an Intruder with 6klbs of fuel. The Intruder approach speed was based on 120 + 2 KIAS per thousand pounds of fuel.

The final descent from 1200 feet commenced at 3nmi from the carrier with a 700 FPM ROD as our closure on the moving carrier was often near 112KIAS. This descent could be made with basic flight instruments but was easier using the Automatic Carrier Landing System (ACLS) approach needles. The needles showed the pilot where he needed to fly the aircraft in both the horizontal and vertical dimensions. To make horizontal corrections, the pilot would fly left or right to line up on the final approach bearing. To make vertical corrections, the pilot added or reduced power to climb or descend to achieve the 3° glidepath for a safe approach to the carrier.

We reached 400 feet MSL at a mile from the carrier where we could normally see the meatball for glidepath information. My role, along with monitoring the approach, was to call out the aircraft's vertical velocity in the descent inside of 3nmi to the carrier. This helped Intruder pilots as the Vertical Speed Indicator (VSI) was not easily readable when looking at the needles or outside the cockpit and responding to the meatball. A variation in our ROD from 700 FPM could result in us being

---

[43] During Case III approaches, the controlling agency would provide a time hack synchronizing all the aircraft.

high or low on the glidepath. Deviations to the value could be expected when trying to get back on the glidepath but the magnitude of the corrections usually remained inside of narrow parameters (600-800 FPM) to maintain a smooth approach.

As aircraft carriers had limited availability for training many new pilots, they were very busy during those times. The FRPs flew with instructor BNs and the FRBNs would fly with instructor pilots or CAT II pilots[44]. The first time I landed on a carrier was aboard CVN-65, the USS Enterprise. We had trained for that moment for nearly a month. My CAT II pilot was a LCDR named Pat. He was a salty character with a big, bushy mustache who oozed naval aviation. Pat was cocky with a quick wit, bright smile and always liked things done well. He didn't suffer fools lightly and could be emotionally passionate when things weren't right.

Monday 5 August Pat and I first flew together for a Field Carrier Landing Practice (FCLP) that we affectionately called 'the bounce pattern'. The FCLP was managed by a trained and experienced Landing Signal Officer (LSO)[45] and his duties were specified in the LSO NATOPS[46]. An LSO was given operational control of 3-4 aircraft in a landing pattern that simulated carrier operations. The LSO was physically positioned near the Fresnel Lens on the left, approach end of the runway. The LSO graded every approach that a pilot performed. His comments were recorded by another LSO and debriefed to the pilot after the flight.

Pat was his own harshest critic. At the conclusion of each landing, he would verbalize how he performed on that pass. At times he would become quite loud and demeaning regarding his own performance. Landing aboard the carrier, 'the boat' as it was called, required precision in controlling the aircraft. Lack of precision resulted in a rough landing or an LSO waveoff in extreme cases. My task during the FCLP pattern was to make the necessary radio calls, keep track of other aircraft in the pattern and support Pat with whatever comments might help. I offered few comments as Pat was covering all that needed to be corrected. Once

---

[44] CAT II pilots were returning to the fleet for a second tour with the A-6E.
[45] What does an LSO do; https://www.youtube.com/watch?v=KY5VtHMz3Ck
[46] The LSO NATOPS can be found here; https://info.publicintelligence.net/LSO-NATOPS-MAY09.pdf

the flight was over, Pat preferred a quick debrief so that he could depart for home.

On 8, 13, 19 and 21 August, Pat and I completed night carrier landing practice in a land-based simulator, the Night Carrier Landing Trainer (NCLT). The NCLT provided an opportunity to practice Case III approach procedures along with night carrier approaches to landing. Each session was 1.5 hours long but seemed much longer; especially that first session.

Contractors that build simulators spend a great deal of time perfecting the hardware and software to closely replicate the feel of the actual aircraft in flight. The visual image needed to closely resemble what the crew would see at the carrier. As this simulator provided a night view of the carrier, it helped the FRPs/FRBNs to prepare for the experience.

For a well-seasoned Intruder pilot like Pat, the NCLT was troublesome. It did not fly like the actual aircraft within 30 seconds of landing. This was when it was most important. This greatly irritated Pat and he felt it was negative training. The instructor encouraged him to do his best.

Pat and I flew day FCLPs again on 9 and 23 August. We flew night FCLPs on 7, 12, 14 and 20 August. After all these flights, Pat was feeling comfortable with his control of the aircraft and ready for the boat. I also flew with younger pilots that would be landing the A-6E on the carrier for the first time. I flew day FCLPs with Mark on the 15th and 22nd, Tom on a night FCLP on the 19th and Lloyd (prospective VA-52 XO) on the 22nd after a hot pump, hot switch with Mark[47].

Tuesday 27 August Pat and I flew to NAS Miramar in San Diego as the USS Enterprise was off the southern California coast. The 28th of August was the big day for our first landings aboard the Enterprise. We had an approximate location and our TACAN provided a bearing and range to the carrier. As a senior pilot, Pat led a four-plane formation to the carrier. As we approached the ship and saw it on the horizon, our excitement grew. I thought, 'Oh my... have a look at that!' Seeing a U.S.

---

[47] Mark, Tom and I would all end up in VA-115 aboard the USS Midway after finishing the FRS.

Navy aircraft carrier at sea is a magical thing. They are big but don't look so when viewed from a distance or from many thousands of feet above it. They generate a wake in the ocean that is visible for a great distance.

The A-6E holding altitude was typically 4000 feet. We entered a circular holding pattern with a diameter of several nautical miles while waiting for the ship to call us down for our landings. One might think this is a time to relax but it isn't. Several planes are flying together in rather close formation at 250 KIAS. Formation flying is something pilots practice a lot when training in a new military aircraft. It is not difficult once you are accustomed to it but requires constant attention as distractions could quickly result in a midair tragedy. We waited for our chance to commence our landings.

The Air Boss made a radio call, "505, Charlie". As it was a training environment and we were instructed to do so, I responded, "505"[48]. That was all that needed to be said. The Air Boss was the senior aviator in the tower controlling flight operations at the carrier. Charlie meant that we were cleared to enter the landing pattern. My response let the Boss know that we had heard his call and would comply.

As we received the call to commence our landing practice, my excitement ramped up again. It was finally happening. All that training was about to be exercised in front of a large audience. All those hours of briefs, simulators, day and night FCLPs prepared us for the next five minutes. We descended aft of the ship down to 800 feet, aligned ourselves in the wake a few nautical miles behind the carrier with our speed at a nominal 300 KIAS. The carrier was much bigger as we flew over in close formation. We wanted to make a good impression with a tight, well-aligned formation. It is common for those onboard the carrier to make subtle judgements about the overhead formations and then the pilots involved if the formation looks sloppy. We broke formation at the bow followed in five second intervals by our wingman.

We slowed the aircraft gradually while descending to 600 feet and performing our landing checklist. Our Flaperon Popup and Antiskid switches were selected off and our hook was up for a touch and go. Pat

---

[48] In the fleet, we would have followed the Boss's direction to enter the pattern without comment.

dropped the gear, lowered the flaps to takeoff position and slowed the aircraft to our approach speed of 132 KIAS. We verified three green lights for the landing gear and checked the fuel quantity. We had six thousand pounds of fuel onboard. Pat started our left-hand turn at the 180 (abeam the ship's stern) with 25 degrees AOB and 300 FPM ROD. As we executed our 180 degrees of turn, the ship moved slightly away from us. We rolled out of our angle of bank aligned with the ships angled landing area at ¾ of a mile at 375 feet. I made the radio call, "505, Intruder Ball, 6.0". The LSO (one from our squadron that has been training us back at our home field) responded, "Roger ball. 25 knots down the angle".

That simple communication did many things. 505 identified the pilot and FRBN in the aircraft. 'Intruder' told the carrier what type of aircraft was about to land to ensure the landing gear cables were set to the correct tension. 'Ball' told the LSO that the meatball was working correctly and that the pilot was using it to guide his flightpath to touchdown. '6.0' told the ship how much fuel the aircraft had in thousands of pounds. Generally, aircrew liked to arrive on the ball with the most fuel that could be carried and yet remain below their maximum gross weight (36,000 pounds for the A-6E Intruder, more for a Tomcat, less for a Hornet). The ship used the fuel weight to again set the arresting gear for tension and helped operations estimate how many takeoffs and landings an aircraft could complete before they need to be refueled (carriers calculated 'Hold-Down Fuel' levels for each aircraft and normally did not launch them again before refueling).

The LSO response confirmed that the meatball was working properly and that we should follow its advice. He also told us how much wind was moving down the landing area. That information told us how quickly our closing velocity was as the headwind slowed down our closure speed. It also told us about how much air disturbance to expect behind the carrier just before landing due to wind-structure interactions. This air disturbance that carrier aviators call 'the burble' could require a lot of very rapid and precise control corrections just before landing to ensure the aircraft did not land hard causing structural damage.

As we approached the ship, Pat put all his practice into action, yet something was different. This was his domain. Pat knew how the meatball and carrier moved. He was a veteran of carrier aviation and it showed. Pat was on top of his game. His stick and throttle movements were quick and deliberate. Pat demonstrated superb control to touchdown. We did a touch and go as briefed for a warmup. We climbed back up to pattern altitude. Pat was elated as the pass was a solid performance. Carrier aviation was clearly the correct vocation for him. He was happy and I was happy for him. But there was no time for congratulations, it was time to stop the climb, look for other aircraft and when appropriate, turn downwind and set up for the next pass.

Everything was done by the numbers. 600 feet. Landing checklist. We were at 5.0 klbs of fuel with a 130KIAS approach speed. We started the turn abeam the carrier's stern. 25 degrees AOB and 300 FPM ROD. For both of us, this was an inside, outside exercise. We were looking at the carrier, monitoring the progress of the turn in relation to the carrier landing area, watching our airspeed, altitude, ROD and looking outside for other airborne traffic. Generally, there should not be any other airborne traffic in this sacred area behind the aircraft carrier. The airborne rescue helicopter was flying on the starboard side of the carrier and well away from our pattern. All other aircraft were well above us. I generally didn't say anything but just watched. If I saw something that Pat was not responding to, I planned to comment if the item was urgent. I didn't want to distract him unnecessarily and didn't need to say anything. Pat was acutely aware of everything that was happening and responded like a master.

We rolled out of our AOB and the carrier was lined up beautifully. "505, Intruder ball, 5.0" was my call. The LSO responded, "Roger ball". Again, Pat was in his groove. His stick and throttle movements were those of a well-seasoned professional. We did a touch and go then heard the LSO radio call, "505, hook down." Pat happily put our arresting hook down. We climbed up to altitude to repeat the pattern. By calling "505, hook down", the LSO told us that he liked what he was seeing, we didn't need more practice and were ready for our first arrestment. Again, we proceeded by the numbers. 600 feet. Landing checklist. 4.0klbs fuel, 128KIAS approach speed. Started the turn at the 180, 300 FPM ROD, 25 degrees AOB, inside, outside. Everything looked sweet. As we rolled out and lined up at 375 feet, Pat responded to movements of the ball like before. Everything was just like the touch and go landings until touchdown.

Once we touched down, Pat applied full power as we caught an arresting cable and life changed. As a new guy not knowing what to expect, I locked my harness retract mechanism to prevent myself from flying forward. We stopped in about a second with our bodies leaning forward from the momentum. Pat, as a salty naval aviation veteran, did not lock his harness and leaned forward at least a foot. With the power forward at full, Pat was ready to fly the aircraft away from the carrier deck but that arresting hook and cable stopped us. Pat pulled the power back to idle, the aircraft was pulled back slightly from the cable that grew slack then he raised the hook. An aircraft director, appropriately named a Yellow Shirt (YS) as he wore a yellow float coat, gave Pat directions where to steer our aircraft[49]. We were going to the catapult for takeoff.

It doesn't take long to get to the catapult and a completed takeoff checklist was required. Checklists are easy to complete when there is ample time and you are in a relaxed state of mind. There is no relaxed state of mind when you have just completed an arrested landing aboard an aircraft carrier. The arrestment scrambles your thinking as it subjects your body to over 3Gs of longitudinal acceleration along the axis of your eyes. It doesn't hurt but it adds confusion to a normal day. I gave a thumbs up to a Green Shirt (GS) on the deck with a 34,000 on his weight board[50]. This correlated with the necessary force that was required for the catapult launch. We quickly completed the checklist as we were taught to never cross into the catapult shuttle before it was done. The checklist was performed as a challenge and reply to ensure that it was done properly. We were ready for the catapult.

The YS directed Pat with increasingly small movements of his nose-gear toward the catapult shuttle. A GS was crouched down low near the nose-gear of the aircraft providing signals to the YS to align the aircraft launch bar on the nose-gear with the shuttle. Pat slowly added power as the launch bar fell into the catapult cradle. The YS turned over

---

[49] The color of the float coat defines a person's job on the carrier; https://www.history.navy.mil/content/dam/nhhc/news-and-events/multimedia%20gallery/Infographics/FINAL_RainbowJerseys_highres_PDF.pdf
[50] There are interesting jobs on the flight deck and this is one of them; https://www.youtube.com/watch?v=dKPWtzAEHGI

direction of the aircraft to the Shooter[51]. He was a naval officer that served as the final safety checker that commanded the launch. The Shooter ensured that everyone on the flight deck agreed to launch the aircraft. One person could stop a launch if they saw something that compromised safety. The GS under the aircraft directed catapult tension then ran away from the aircraft. His job looked like the most dangerous aboard the carrier. The Shooter scanned the deck for any concerns then directed Pat to add full power. A holdback fitting kept our aircraft in position until the catapult fired.

We completed final engine checks of hydraulic and oil pressure, Exhaust Gas Temperature (EGT) 605-677° C, Revolutions Per Minute (RPM) 94-101.1% and Fuel Flow 6200-8500pph. Pat performed a control wipeout to confirm no binding, unusual movements or malfunctions. The control and engine checks took about 10 seconds. Once completed to our satisfaction, Pat saluted the Shooter. The Shooter checked around the deck again for any signs of danger then touched the deck and pointed forward to command our launch. A GS in the catwalk pressed a button that fired the catapult and we raced down the bow of the Enterprise[52]. The time from salute to launch was about eight seconds.

The first thing I felt on the catapult launch was the breaking of the holdback fitting. The fitting was a metal barbell-shaped device that was about 4 inches long. The fitting was strong enough to hold the aircraft to the carrier with its power at military but would break when the catapult fired[53]. When it broke, there is a noticeable snap and the acceleration ramped up quickly to over 3 longitudinal Gs along the axis of our eyes. It was like an amusement ride with lots of noise and acceleration then smooooooth as we left the deck and were flying. Pat raised the gear and lowered his hook out of habit as we checked our airspeed and climbed to 600 feet. We looked for other traffic in the pattern and sequenced in behind them. We repeated the arrested landing sequence then refueled before completing more catapults and arrested landings.

---

[51] The role of the Shooter was a great responsibility; https://www.youtube.com/watch?v=YgcMP-PyfTg
[52] An amazing view of the work of the catapult deck edge operator; https://www.youtube.com/watch?v=g2XgPbc9tbI
[53] As the fitting broke, half stayed in the holdback fitting, half remained locked into the nose-gear for removal after landing.

We finished our daytime work with the two touch and goes, six day arrested landings and five-day catapults. Pat jumped out of the jet and an instructor LSO, Rich, jumped in with me for a day catapult and arrested landing. As an experienced LSO, Rich was also a very smooth pilot. Again, everything was flown by the numbers with checklists and cross checks. Once back on deck, Pat and I waited for the night flight operations.

The carrier flying during the day was rather straightforward for Pat, but the night could be more challenging. We manned the jet on the flight deck as it was getting dark. The INS alignment worked differently as the carrier passed position information via a datalink. We taxied to the catapult under the control of a YS who used yellow light wands to direct our movement. I confirmed the aircraft weight of 45,000lbs with the GS[54]. We were over gross weight for landing but would burn off the extra fuel in holding and on the approach for the landing. Pat applied full power when directed then proceeded with engine and flight control checks. When ready, he turned on our tail anticollision light indicating that we were ready for the catapult.

The catapult felt as during the day, but we climbed away from the water on instruments rather than looking out the windscreen. Outside the aircraft was pitch black and provided no reference for visual flying. The carrier-based approach controller provided our Case III holding instructions. We climbed to our holding altitude and established our racetrack pattern to commence our approach on time. Our training had prepared us to complete night Case III approaches and carrier landings as if they were routine.

We pushed on time heading toward the carrier, initiated our 4000 FPM ROD and focused on our instruments in the descent. We decreased our ROD to 2000 FPM at Platform and continued inbound. We leveled at 1200 feet MSL, transitioned to the landing configuration at 10nmi and slowed to our approach speed of 132KIAS. The ACLS needles assisted us in maintaining our lateral position along the Final Bearing. We

---

[54] The Intruder maximum gross weight for catapult takeoff was 60,400lbs. Shots at that weight were quite powerful.

commenced our descent at 3nmi with the 700 FPM ROD. Like in the simulator, I called our VSI values.

As I wasn't controlling the aircraft, I was able to glance more often out the windscreen at the carrier. It looked a lot like the one in the NCLT simulator. The lights of the carrier against the dark ocean were beautiful to see. The yellow meatball and green datum lights, the dim white lights outlining the landing area and the red vertical lights that extended on the stern from the deck to the waterline were stunning. The internet has various sources to appreciate the experience of a night carrier landing[55].

We made the ball call and trapped on the deck in the soft glow of lights. Occasionally, for various reasons, an aircraft doesn't catch a cable resulting in a bolter. In that situation, the pilot applied full power, performed an instrument climb away from the water, followed the instructions of the approach controller regarding headings and altitudes and set up for another approach. We had six-night catapults and six arrested landings with one bolter. I felt comfortable with our night landings due to the extensive training we received. Pat and I spent the night aboard the ship to complete more landings the following day.

Thursday 29 August Pat and I completed four, day catapults and landings before flying back to NAS Miramar. We refueled the jet and flew on to NAS Whidbey Island. It felt great to have successfully completed the carrier operations portion of the syllabus. Pat was happy as he was through the toughest part of the syllabus for pilots. He was anxious to move on and get to his next fleet squadron.

Tuesday 10 September - Saturday 28 September Not much time had passed since returning from the USS Enterprise before we headed off to NAS El Centro for a visual bombing detachment. El Centro, in the southcentral corner of California, enjoys crystal blue skies for most of the year in a dry climate. Two groups of students, our class and the one behind us, came to El Centro to make the training time more efficient.

---

[55] An S-3 with split day/night view of carrier landings; https://www.youtube.com/watch?v=g5zgJFYFIRg; Carriers at War, Smithsonian Channel, https://www.youtube.com/watch?v=rmT_k2V2uTE

FRPs and FRBNs learned various aspects of visual bomb delivery including the 40° dive bomb pattern, level laydowns and pop-ups. Most of the weapons used in this training were 25lb, inert MK76 bombs. There was a multiplane, coordinated strike that included dropping live MK82 bombs on a government target in the middle of the California desert. Finally, ACM and tactics were also taught on this detachment. My first flight from El Centro was with Axel practicing 40° dive bombing and my last was with Tuck, a FRP from the class behind.

During the middle of the detachment, I walked into the squadron ready room to await my preflight brief. Another brief was in progress and my future Eagle squadron mate, callsign Mildew, was the Squadron Duty Officer (SDO). He took one look at me and said in a loud voice, "Marblehead". The aircrew briefing the flight stopped, looked at me and seemingly simultaneously started laughing. "Yes, that's it" one of them announced. "That is your new callsign" and from that moment on, I had a new name. The worst thing that an aviator can do is to try and disavow a callsign. The more the battle grows, the more the callsign sticks. I earned several variations to 'Marblehead' over the years.

Instructor LCDR pilot, Lee and I completed a two-hour, night flight completing multiple 40° dive bomb deliveries with a division of four aircraft. This flight was a real eye opener! Lee had skill and demeanor that bred confidence. The partially illuminated target was on a training range but screaming down at the ground at night required accuracy, discipline and consistent adherence to flying the numbers, releasing the bomb at the right altitude and getting that nose away from the rapidly upcoming terrain. That required a 5G pullup adding more physiological challenge that was already present in a night mission. Keeping track of the other three aircraft required persistence attention. Lee stepped the aircraft into the attack mode of the mission computer placing a reticle 20 mils below the aircraft datum line on his gunsight. At the correct spot on a cone around the target, he overbanked the aircraft rolling inverted to pull the nose down to meet the imaginary 40° cone. I waited for the aircraft to be pointing at the ground before activating the onboard LASER then called out dive angles and altitudes as Lee was focused on flying the aircraft to position the reticle on the target. There wasn't flight information on the gunsight like it is with the Heads-Up

Display (HUD) in modern aircraft. The BN calls provided the necessary situational awareness in the dive.

There was a computer mode in the Intruder that allowed the pilot to designate near the target with the BN correcting the aircraft aimpoint using the FLIR crosshair. This was called a BN Handoff and could produce bullseye hits. Near our last bomb drop, Lee asked if I'd like to try one. The challenge was that the BN had less than 5 seconds to designate and track the target in a 40° dive before the bomb came off and the pullup started. I agreed and placed my face up against the hood after we rolled into the 40° dive. I turned on the LASER, Lee designated the target, I saw his designate near but off the target then slewed my FLIR reticle over the target and started tracking it. Lee pulled the Commit trigger and followed steering commands enabling the computer to release the bomb at the appropriate spot in the sky. The bomb came off and I continued tracking the target as he executed the 5G pullup away from the ground. We nearly scored a bullseye and were very excited about the result. We joined up with our other wingmen and returned to El Centro for the debrief. It had been a demanding but very rewarding night of flying.

The final flight of the Visual Weapons detachment was flown with FRP, Tuck. He was a happy go lucky guy that had a perpetual smile on his face. He looked more like an accountant or lawyer than a pilot. Looks were deceiving as he was a great Intruder pilot as well. The ordnance release was a load of ten, high explosive MK82 (500 lb) bombs during a multi-aircraft, coordinated strike against a target in the government controlled, Chocolate Mountains[56]. The weapons came off as advertised, causing a big shake on the aircraft and exploded on the target.

Tuesday 22 October I flew a day bombing flight with instructor pilot, Jeff. The mission was to practice bombing on a moving vehicle. The Boardman target range had a remote-control, full size dune buggy that raced around the target area and we had the opportunity to attack it. The Intruder radar had an Automatic Moving Target Indicator (AMTI) mode that erased all ground returns except those that were moving. The

---

[56] The Chocolate mountains;
https://www.wikiwand.com/en/Chocolate_Mountain_Aerial_Gunnery_Range#/overvie
w

BN prosecuted an attack on those items that were moving by slewing the radar cursors up to the blip, selecting a radar track mode, stepping the computer into attack and releasing a weapon. If the FLIR was functional, lasing the target and a FLIR handoff added to the bomb's accuracy. We dropped a dozen bombs on the dune buggy with good hits. Once the bomb came off our aircraft, Jeff called "bomb" over the Boardman control frequency. The dune buggy controller would abruptly change its course to prevent it from getting hit. Despite our bombs being inert and only weighing 25 lbs, if they hit something traveling at 500mph, it usually stopped working. Previously the buggy wasn't turned before the bomb impact and it suffered the consequences.

Wednesday 23 October Today's flight was unlike others by design. Both FRPs and FRBNs were taught courses on aviation intelligence, electronic warfare and nuclear weapon delivery. As the prime naval asset for nuclear delivery, this course was critical for Intruder crews. We learned about the nuclear weapons that the Intruder could carry. The briefings were all classified Secret and held in secure spaces. Every Intruder crew flew to an unfamiliar target to deliver a simulated nuclear weapon with an instructor verifying the successful completion of the mission. The aircraft was pre-flighted as normal with the addition of a simulated nuclear weapon mounted on the centerline station of the aircraft. While this was a training exercise for me, it was also a big training exercise for the ordnance men. They had to load the weapon according to a very strict set of rules and protocols. The flight launched from NAS Whidbey Island, entered the visual low level, VR-1352 and delivered the weapon on the B-20 bombing range adjacent to NAS Fallon, NV.

The delivery tactic was a high loft that threw the bomb toward the target in a steep climbing maneuver that allowed the Intruder to complete an Immelmann away from the nuclear blast. I was nervous as we approached the target after having spent a great deal of time studying the area. The weapon had to be delivered on the first pass at a specific time for a successful completion of the flight. I was able to keep the navigation system updated and found the target early. We added full power and my instructor pilot followed command steering to release the weapon. The weapon scored a reasonable hit on the target. A bullseye is not required for nuclear weapons.

We landed at NAS Fallon, refueled and flew back to Whidbey. I was relieved to complete this additional big milestone in my training. No Intruder crew has ever released a live nuclear weapon, but all the crews were trained in how to do so if required. We were in the middle of the Cold War with the Union of Soviet Socialist Republics (USSR) and we didn't really know what the future might hold for us.

Thursday 7 / Friday 15 November I flew the second day/night low level, the IR-344, with a Navy Reserve pilot, Dave. He was a relaxed guy with an easy-going demeanor. The day flight proceeded smoothly with a weapon release at Boardman and a high altitude return to Whidbey. We briefed the night flight, took off and flew southwest past Seattle to the route entry point along the coast west of Aberdeen, WA. We descended over the water on a southwesterly heading for 36nmi then made a left turn to cross the beach near Oysterville, WA. As we approached the coast at 1000 feet above the water, we saw car lights going down the road. It was exciting to see. Our dialogue on the low level was natural and productive. The bombing run at Boardman proceeded as briefed with a reasonable score on the target. We returned to Whidbey uneventfully at high altitude.

Saturday 16 November The IR-346 day/night low level combination was flown on a single day with another Navy Reservist, Roy. This was normally not done for crew rest reasons but the squadron executives were anxious to get me to Japan. I felt comfortable with the scheduling as did Roy. Had either of us felt tired or not up to a second flight that day, we could have cancelled the night flight. We both felt well-rested, so we proceeded. We took off from Whidbey and climbed up to FL290 for a transit south to Portland, OR at 420 KGS. We turned southwest to intercept the coast at Newport. We descended along the coastline for 34nmi to 15,000 feet MSL then turned over the water for a further descent to 1000 feet AGL.

Our speed was 420 KGS when we recrossed the coast heading east near Reedsport and started our SRTC dialogue. The first leg paralleled the Umpqua River for several minutes. We stayed on course as there was a cable rising to 800 feet AGL a few nmi left of course. We crossed Interstate Five near Cottage Grove, started climbing into the mountains and entered instrument conditions due to snow. We made a ninety degree turn to the left at the dam at Hills Creek Lake. We couldn't see the ground visually yet felt comfortable as the SRTC was working well and our crew coordination was in synch. We flew over our turn point at the Cougar Reservoir Dam and proceeded north. The snow continued as we passed Detroit and Timothy Lakes with terrain rising to nearly 6000 feet MSL.

We turned to an easterly heading and descended out of the Cascade Mountains and the instrument conditions. We avoided the town of Maupin due to heavy crop dusting in the area and proceeded toward the Boardman bombing range after the Condon Ranch turn point. Our radio check-in with the bombing range proceeded smoothly and we released our MK76 bomb with another good result. As the weather was poor in the mountains, we returned at high altitude back to Whidbey.

The flight had been a good workout. Roy and I worked well together. We debriefed the flight, had something to eat and briefed the night flight. As everything was fresh in our minds, the brief was short. The flight proceeded very much like the day flight with one major, memorable difference. The poor weather and snow were still in the mountains during the low level requiring our refined SRTC flying skills.

Not long into the low level, Roy and I looked up to see Saint Elmo's Fire[57] dancing radially around the aircraft refueling probe in a thin, ring with a diameter of about eighteen inches. Neither of us had ever seen anything like it. "Wow, look at that", "That is amazing" were comments uttered. We stared at the probe and blue dancing electricity for about 10 seconds. We were mesmerized by the image. Suddenly, we both realized that we were at low altitude, at night in the mountains in a military jet that required manual inputs to ensure our safety. We ignored the probe, got back to our screens and dialogue about the terrain out in front of us. Fortunately, nothing sinister occurred during our temporary fixation and we were able to safely resume our duties. Later, in a less stressful portion of the flight, we discussed the amazing nature of the St. Elmo's fire. Most Intruder crews saw it on the windscreens when flying

---

[57] Plasma discharge phenomena called St. Elmo's Fire; https://www.youtube.com/watch?v=vcKCmZelQA0 and https://science.howstuffworks.com/nature/climate-weather/atmospheric/st-elmo-fire.htm

in Instrument Meteorological Conditions (IMC), but few had ever seen it dance on the probe like we did that night! It created a permanent, vivid memory.

Tuesday 26 November My Intruder NATOPS check ride occurred in the A-6E WST with a junior FRP. The instructor pilot at the control console introduced simulated aircraft system failures to verify my ability to work with the junior pilot and safely solve the problems. The simulated mission was an aircraft carrier launch to an IR-344 instrument low level followed by a bomb release at Boardman and a TACAN approach at Yakima airport. The mission went well and provided the student pilot insight for his NATOPS check. With the graduation exercise completed, I was cleared to join the VA-115 Eagles aboard the USS Midway. I started the FRS in March and completed in November 1985 after 51 flights in the Intruder and 11 flights in the TC-4C training aircraft.

# Chapter 5

## VA-115 Eagles, USS Midway, NAF Atsugi and the Far East

*Seconds after our bomb release, an F-5 fighter swooped down on us from above right. Mike put on a hard 6.5G turn into him hoping for an overshoot. That turn lasted about 10 seconds then 6.5Gs back to the left for about 10 seconds. The F-5 was much more agile than we were and nestled in behind us for the kill.*

To maintain global peace, the United States forward deploys military personnel and equipment to geographical areas of concern. They do so at great financial cost to avoid the much greater costs of war. Those costs include the loss of human life, displacement of people from their homes, destruction of productive lives, loss of civilization, infrastructure, peaceful existence and hope for a better life. The ships of the USS Midway battle group, the aircraft, the officers and sailors along with large groups of support personnel represented a substantial investment in peace. Those involved were proud of their contributions and enjoyed the overseas experience despite the challenges.

It isn't always easy living in foreign countries. We missed our families back in the United States and the ease with which tasks could be completed. At times, simple jobs in the USA could be very challenging overseas due to infrastructure or cultural differences. We had to be innovative to get the job done. It added adventure to the experience (and at times, frustration). This permanent placement overseas developed a unique culture among those involved. A VA-115 squadron patch exemplified the feeling; 'Permanent WESTPAC'ers[58]. We aren't tourists. We live here'. The patch was partially a social commentary about the stateside-based aircraft carriers and crews in this era that experienced an overseas deployment for six months every few years[59]. The Eagles flew from the USS Midway, that was homeported in Yokosuka, Japan about 30 miles south of central Tokyo. Their primary, land-based airbase was

---

[58] Western Pacific (WESTPAC) was considered west of the International Date line to the Persian Gulf.
[59] Since September 11, 2001, military personnel globally have deployed more frequently to combat terrorism.

the Naval Air Facility (NAF) Atsugi[60], located 24 miles southwest of Tokyo. During the 1986-1988 timeframe, the Eagles had a middle east deployment, land-based deployments to South Korea, the Philippines, Okinawa, NAF Misawa[61] and the USA locations of NAS Whidbey Island and NAS Fallon, Nevada.

Serving aboard sea-going, military ships can be one of the most memorable experiences of a person's life. Those that have experienced it, endured many challenges but often look back fondly to their days aboard ship. The USS Midway became the first and only permanently deployed U.S. aircraft carrier from 5 October 1973. The Japanese tradesmen that worked on the USS Midway while she was in Yokosuka took great pride in their efforts. She was their aircraft carrier as well. They were honored to be working on the ship. Serving aboard the Midway provided many amazing memories. Before sharing others, one captures the spirit and honor of the duty.

The USS Midway was in the Arabian Gulf during a six-month deployment in December 1987 helping keep the peace during a politically tense time[62]. The Iranians were attacking Kuwaiti oil tankers as they passed through the Persian Gulf, the Strait of Hormuz and into the Gulf of Oman. The USA initiated Operation Earnest Will[63] in response. Earnest Will directed the U.S. military to escort reflagged Kuwaiti tankers through these dangerous waters to ensure their safety. VA-115's mission was to attack Silkworm missile sites on Iranian soil if they proved to be a threat to shipping. As Midway is an oil-fired ship, we needed to replenish our fuel supplies every few days to remain combat ready[64].

---

[60] NAF Atsugi; https://cnrj.cnic.navy.mil/Installations/NAF-Atsugi/
[61] NAF Misawa resides on the US Air Force Misawa Air Base; https://cnrj.cnic.navy.mil/Installations/NAF-Misawa/
[62] https://commons.wikimedia.org/wiki/File:Battle_Group_Alpha_(Midway,_Iowa)_under way,_1987.jpg
[63] Operation Earnest Will escorted Kuwaiti tankers https://www.globalsecurity.org/military/ops/earnest_will.htm
[64] Underway replenishment; https://en.wikipedia.org/wiki/Underway_replenishment; https://fas.org/man/dod-101/sys/ship/unrep.htm

During one of those replenishments, I was on the flight deck looking across at the supply ship, U.S.S. Cimarron (AO-177). The USS Midway and Cimarron were steaming parallel to one another with about 100 feet of separation between them. It reminded me of being a midshipman in 1979 aboard the USS Sacramento while refueling the USS Coral Sea. Over USS Midway's 1MC onboard loudspeaker system was Lee Greenwood singing God Bless the USA[65]. The contributions of the United States were very evident at that moment. Two American ships, full of sailors steaming a long way from home with a great love for their country, were right before me. The folks behind the scenes that contributed to our presence there were too numerous to count. Ultimately, the U.S. taxpayers paid to have us on station that day and deserved to see the moment. It was precious yet seemingly impossible to share.

My last flight at VA-128 was on 20 November. I took annual leave in Colorado to see my family then flew from Denver to Seattle on the morning of 13 December to await a ten-hour flight to Tokyo Narita airport. I secured military transportation to the USS Midway and slept in bunkroom six[66]. The next day, a van transported me 1.5-hours to NAF Atsugi dropping me off at the BOQ. The buildings were nearly new and everything was meticulously clean.

I called one our pilots, Beav[67], from my class at VA-128. He explained the squadron organization[68] and shared many sea stories. We met up with Eagle BNs Bob, Bart, and Komrade along with KO. We all had a great dinner at the Atsugi Officer's Club that was positioned across the street from our BOQ. I went to sleep but woke ready for the day at three am. Jet lag was affecting my biological clock.

Monday 16 December Beav and I arrived at the VA-115 hangar by seven am. The other aircrew at the squadron were welcoming. We participated in an All-Hands flight line Foreign Object Damage (FOD)

---

[65] Lee Greenwood's God Bless the USA;
https://www.youtube.com/watch?v=pwEcz9nABNg
[66] Bunkroom Six was a VA-115 Junior Officer (JO) Stateroom. I later resided in the adjacent, bunkroom eight.
[67] Beav's callsign was given as he looked like a grown-up Jerry Mathers from the Leave it to Beaver show.
[68] A squadron organizational chart;
https://commons.wikimedia.org/wiki/File:Navy_Squadron_Organization.png

walkdown. These walkdowns became part of everyone's life. Jet engines are easily damaged by anything solid going down the air intake especially anything made of metal. We walked in a straight row picking up anything that we could see. Generally, the flight line was clean but we still found all sorts of strange debris.

I met our Commanding Officer (CO), CDR Dusty Rhoades, a pilot with a Hornet flight test background. I'd only heard great things about him, and they appeared to be true. I became the new AQ[69] Branch Officer and Coffee Mess Officer (CMO)[70]. Our XO, CDR Paul Cash (another pilot), invited me to Christmas dinner with his family. That was really appreciated.

Tuesday 17 December We had an All-Officers Meeting (AOM) in the morning. The highlight was a brief on the Philippines detachment and the Cope Thunder exercise coming up in January. The detachment was to be at NAS Cubi Point, adjacent to Subic Bay Naval Base and the neighboring town of Olongapo. The flying sounded great with the airwing working together on attack/fighter strikes opposed by Air Force fighters launched from nearby Clark AFB. I was given my NATOPS manual and Squadron Operating Procedures (SOP) to review and complete open book tests. I read through the squadron SOP, took notes and completed the tests.

Wednesday 18 December I turned in my NATOPS and SOP tests and started the process of learning the responsibilities of an AQ Branch Officer.

Thursday 19 December We had a Squadron awards ceremony in the morning. Many of our enlisted men received awards for exceptional service. We had a Squadron Christmas party later that night. I met our new Intel Officer, Vern. He went to school at CU Boulder but I'd never met him as it was a big school. Vern seemed like a nice guy but a

---

[69] An AQ was an Aviation Fire Control Technician. They maintained the Intruder's radar, IMU, INS, FLIR, Mission Computer and weapon station wiring.
[70] The CMO coordinated Hail and Farewell parties, ordered plaques, patches, coffee cups and memorabilia.

bookworm[71]. We helped him loosen up and get accustomed to our pack of flyers.

Tuesday 24 December The Squadron had a 'Green Light'[72] at our CO's house at 7pm. There was a lot of good food, drink and camaraderie. Afterwards, many of us went over to the O' Club and played a game with a cup and five dice[73] while sharing a beer. Kenny-San and Charlie-San were our mature-aged, Japanese bartenders who became like family over time. Upon arrival back at the BOQ, we found Christmas stockings full of candy and presents courtesy of our XO.

Wednesday 25 December Christmas Day started with an afternoon meal with the XO's family and pilots Beav, Tom and his wife, Beth. The meal was amazing with turkey and oyster dressing. After the meal, we played Trivial Pursuit then went outside to play with a new football. We started a game and was joined by the CO's guests from next door. The game became the XO's guests vs. the CO's guests. XO's guests won 10-7 with fun had by all! We returned to the XO's home for a dessert. It was a great day with our new Eagle family.

Friday, 27 December Work restarted after the holidays. Our squadron was suffering with less reliable radar, FLIR, computer and inertial navigation systems that were maintained by the AQs. Without these systems, the aircraft was a limited bomber and more like a big A-4 Skyhawk.

Monday 30 December Pilot Waldo and I shared my first flight in Japan. We briefed a standard familiarization flight around the local area. After maintenance control, we put on our flight gear, opened the door to

---

[71] Vern became a great friend and encourager while at VA-115. He is a super nice guy and loved by everyone.

[72] A 'Green Light' is a great Naval tradition. A group of people get together with drinks and food and call someone. When answered, the person calling says 'Green light' and hangs up the phone. The group shows up at the door of the person called, 5-10 minutes later and invite themselves in.

[73] This game is played at many military clubs around the world with variations. The goal was to roll the highest score on the die with up to three rolls and ones being wild. The winner of the first round orders the drinks. The loser of the last round pays for the drinks for those playing. Losing a game with many players can be expensive.

the small operations office to let the SDO know that we were walking to our jet.

All the aircrew stood the SDO duty once familiar with the squadron operations. The job was to execute the daily flight schedule. The SDO often had many things on his mind as he juggled multiple problems. There were often maintenance or operations issues that would pop up requiring changing schedules, crews or jets. It could be a hectic job when the phone was ringing, maintenance control was calling on their direct line and aircraft were calling in on the radio. If another officer was in the room using the radio to convey emergency procedures for an aircraft having a problem, it was a very active location.

Waldo and I pre-flighted the jet and started the engines. Our procedures and those of the maintenance crews were like the FRS. The air traffic controllers that provided our flight, taxi and takeoff clearances were Japanese nationals. While their English skills were excellent, at times their accent was difficult to understand requiring them to repeat their directions. We took off from runway 19 and flew a southerly heading. Once outside the airfield boundary, there was a dense array of homes and businesses. There wasn't an open area of land available to build a house. The view from the air showed how crowded it was around the Kanto Plain[74]. There were 37 million people living in the greater Kanto Plain in 1985[75]. Waldo and I practiced high-altitude bombing maneuvers without any ordnance.

---

[74] The Kanto Plain includes the greater Tokyo area and is slightly larger than the state of Massachusetts in 2021. https://en.wikipedia.org/wiki/Kantō_Plain
[75] Population estimate for 1985; https://en.wikipedia.org/wiki/Kantō_region

Climbing aboard an Intruder while with VA-115. We flew whatever jet was available, not just one with our name on the side.

Tuesday, 31 December We worked half-day as it was New Year's Eve. The squadron officers had a progressive dinner party. We had hors d'oeuvres and drinks at the Operations Officer's (OPSO) house, a big, healthy salad at the CO's house, the main course of turkey, lamb and oyster stuffing at the XO's house and dessert at the Safety Officer's house. We moved over to the Maintenance Officer's (MO) house for dancing and champagne at midnight. It was easily the best New Year's Eve party I'd ever been to.

Friday 3 January 1986 Eagle pilot, callsign Pru and I shared my second flight from Atsugi. He was a senior Lieutenant (LT), had years of flying around WESTPAC and was the Pilot NATOPS Officer for the squadron. Pru and I spent most of a 1.7-hour flight in a KA-6D tanker in VFR conditions without talking to any controlling agencies. We took off from runway 19 and cancelled our flight following with the control tower once well south. We flew very close to Mount Fuji[76] and looked down into the crater. I was very excited to be flying around Japan and

---

[76] Mount Fuji rises to 12,388 feet above sea level and provided many majestic photo opportunities.

considered it a priceless experience. We proceeded low level west of Fuji flying wherever we wanted as there were no designated routes. We avoided towns and cities out of courtesy and respect. We covered a large amount of terrain all while at a few hundred feet above the ground.

We ended up north of Atsugi, checked in through Yokota AFB approach controllers who handed us over to Atsugi Tower for an uneventful landing. That night, a group of Eagle aircrew headed over to the O' Club for dinner. We loved being in Atsugi.

Monday 6 January Our Carrier Air Group (CAG) LSO, callsign Duck and I shared a flight. He was an easy-going guy with a great sense of humor and a big smile. I planned a low-level route using a chart to track the most interesting terrain in western Honshu.

Tuesday 7 January VA-115 had a Safety Standdown along with our AOM. There were more briefs about our upcoming detachment to the Cope Thunder exercise in the Philippines. The training provided a realistic combat environment where inert bombs were dropped on targets while enemy fighters tried to stop the attack. It brought both Navy and Air Force aircraft together in the exercise. Everyone was very excited to participate and get away to the warm Philippine winter.

Monday 13 January Pru and I shared a 4.5-hour flight in an Intruder to NAS Cubi Point in the Philippines. We conducted a preflight brief with the crew of our KA-6D tanker wingman. The weather forecast predicted fine weather along the route and Cubi was balmy.

We took off with the KA-6D on our wing and climbed up to FL280. Three hundred miles south of Atsugi, we passed the lead to the tanker and started inflight refueling. We filled our tanks and proceeded to Cubi while the tanker returned to Atsugi.

As we approached Cubi, we listened to their Automatic Terminal Information Service (ATIS)[77] and hear the weather was fine as forecast. Cubi was a primary WESTPAC hometown for naval aviation, so it was important to look good for our arrival. Other aviators and maintenance

---

[77] ATIS provides an overview of the winds, clouds, altimeter setting and expected approach and runway in use.

personnel would watch the aircraft arrive from the ground. We requested and were granted a carrier break. Pru descended to 800 feet AGL and pushed his power up to get 450 KTAS. We broke at the numbers and Pru pulled on 6.5 Gs as we slowed in our left hand turn over the water. We quickly completed our checklist and landed uneventfully. As we cleared the runway, the canopy came open and the humidity hit me. "Welcome to Cubi", Pru exclaimed. I was really impressed as it reminded me of Hawaii.

In the late afternoon, we went to the Cubi O' Club[78] for drinks on the verandah overlooking the airfield and pier. It was an amazing view and the place to be! We sat and watched jets screaming into the carrier break for landing. As we enjoyed the view, the sun set and put on a glorious show. The fruit bats came out and covered the sky. Sunset was not a good time to come into the break as hitting a bat was commonplace.

Tuesday 14 January My first flight in Cope Thunder was also my first with Ed. He was a charismatic, confident and friendly lieutenant that loved to fly to the edge of the aircraft's envelope. He hoped to be selected for the Navy Blue Angels flight demonstration team so he took his flying seriously. We briefed with a big room full of people who were connected by a phone link to the Air Force fighter adversary pilots flying from Clark AFB. Our CO was the lead pilot for our division of four aircraft. The standard route for Cope Thunder had the attack aircraft and their fighter escorts entering a low-level route about 8 minutes flying time north[79] of the Crow Valley target complex[80] and proceeding south at 420KTAS. The speed was increased to 450KGS for the final inbound leg to the target. Egress was to the west at the maximum speed the jet could produce.

We entered the low level at 200 feet AGL with a pair of F-4 Phantoms within a couple hundred yards from our wingtips. Having

---

[78] The story behind the infamous Cubi O' Club; https://www.youtube.com/watch?v=xqYzyIeOYJ8. The Cubi O' Club was eventually moved to Pensacola, Florida; https://www.youtube.com/watch?v=SxN6RlDgulA
[79] We entered the low level near Dagupan City, 16°02′N 120°20′E.
[80] Crow Valley Target 15°15′28″N 120°22′30″E
https://en.wikipedia.org/wiki/Colonel_Ernesto_Rabina_Air_Base

fighter support so close was a nice feeling. The Intruder didn't generally carry any defensive weapons against enemy fighters as the five weapon stations were usually loaded with bombs or fuel tanks. Having aircraft to protect us didn't alleviate our need to keep a sharp lookout for enemy fighters. As we proceeded south, our fighter escorts got a radar contact with the inbound enemy adversaries. They added power and accelerated out in front of us to engage the enemy. We continued our inbound course hoping that our fighters dominated the engagement. If the enemy got through our fighters, we would turn towards them until intercept then resume to the target.

The terrain from the route entry point to Wild Horse Creek was flat with mountains that rose a couple thousand feet off to our right. Off to the left was flat rice patties. We could easily see fighters over the rice paddies, but they could use the mountains to mask themselves while they looked down on us. We spent most of our time looking toward the mountains. At our closure speeds, the enemy could be no-where in sight then on top of us in the matter of 10 seconds.

As we turned 30° to the right to commence our attack run, Ed added power to get 450KGS. With the large radome on the front of the aircraft, the Intruder was slow to accelerate, especially with a load of heavy bombs. It was a great bomb truck but not a quick and agile bomb truck. As we approached the target area our eyes were outside looking for our specific target as well as still scanning for the enemy. Target area study helped us recognize key landmarks to enable attack on the correct target. The time from target area acquisition to recognition to being lined up on the desired target before bomb release was often a mere 30 seconds. Post-release aircraft maneuvering depended on the weapons released. Retarded bombs with extended fins or parachutes that rapidly decelerated behind the aircraft, allowed egress straight ahead to escape the weapon fragmentation pattern. An aerodynamic, slick bomb required an accelerated turning maneuver of 4+ Gs to safely escape.

On this flight, Ed designated the target visually using his gunsight then put on an accelerated turn to the right to escape the frag pattern. The egress heading away from Crow Valley was due west with rising terrain to 5000 feet and large canyons that extended for 30 miles to the sea. Ed put on a big 6.5G turn at the target and didn't stop cranking the aircraft back and forth until we reached the ocean. That was a tiring, few minutes of flying. Once 'Feet Wet', meaning over water, we slowed down and climbed up to 5000 feet MSL while heading south to return to NAS Cubi for a carrier break. Once over water, we discussed the attack and target. We didn't see any enemy fighters and dropped our weapons on the correct target.

Once in the debrief, we received news that we had survived the adversaries, delivered our ordnance very close to the target and looked good with defensive maneuvers as we exited the area. We were happy with that feedback.

The XO and I shared a dusk/night bombing flight at Tabones Island, not far offshore from NAS Cubi. As we arrived, there were small banca fishing boats around the island and our target rock adjacent to the island. We made a low pass to let them know we were there for MK76 bombing. We were practicing radar to FLIR target handoffs on the rock. I acquired and designated the target on my radar and waited for a clear image to appear on the FLIR. The high humidity in the Philippines made the FLIR less useful than in dry climates. The targeting worked well with good weapon impacts.

As the XO and I were dropping our 25lb practice bombs, we were careful about where we were and at what altitude. There were large mountains immediately to the east of the target. We stayed over the water to the west to avoid them. As we were over water at night at rather low altitude, we carefully monitored the aircraft's instruments. Descending turns at night at low altitude over the water were a prohibited maneuver as it was possible to forget about the descent and fly into the water. An Eagle crew skipped off a wave after forgetting to keep an eye on their altimeter in a descending turn about a year before my arrival. They landed at NAS Cubi with their engines pushed back into the airframe structure, the centerline tank ripped off the jet and signs of seaweed in the intakes. They were lucky to survive but the aircraft had a long, strange history following the event[81]. This reinforced our need for vigilance when flying at low altitude over the water at night. Once all our bombs were used, the XO and I returned to NAS Cubi for landing practice.

---

[81] The troubled aircraft; https://www.smithsonianmag.com/air-space-magazine/mystery-christine-180977829/

Wednesday 15 January Ed and I shared a second Cope Thunder mission. Our target at Crow Valley was a simulated Surface to Air Missile (SAM) site. Our bomb load was ten Snakeye (retarded), inert MK82 (500lb) bombs[82]. These bombs weren't explosive but gave us a great feel for the aircraft maneuvering limitations with the 5000 extra pounds on the wings. The mission was like the first but more comfortable with clearer expectations. We started our low level as on the first mission and flew south towards the Crow Valley target area. We had fighters escorting us once again. We loved that feeling of protection especially as we had no air-to-air radar to see inbound bandits. The escorts were stepped up above us and closer to the mountain ridgelines to respond quicker to the inbound adversaries. We were at 200 feet AGL in the rice paddies cruising along at 420KGS.

We made our right turn before the target, accelerated to 450KGS and started looking for the SAM site. As we approached 3.5nmi from the target, we executed a right hand, 4G pop-up, rolling upside down to pull the nose down onto the target. The higher altitude provided a greater vantage point for finding the target with a shallow dive to release our weapons. We spotted the target in our turn, Ed designated and the computer commanded weapon release. The wings shook as the bombs came off. We continued our descent back down 200 feet AGL as our speed grew to 480KGS. Our weight and drag had dramatically reduced during the release. Ed turned west toward our egress route and the sea. He once again jinked the aircraft back and forth through the mountainous valleys with pulls up to 6.5Gs. As we arrived at the coast, Ed pulled back the power and climbed us to 5000 feet for the return to Cubi.

Saturday 18 January An afternoon mission was with a squadron department head, Mike. He was a LCDR, respected for his piloting skills and was a nice guy. Our division of aircrew had previous Cope Thunder experience so we knew what to expect. Unfortunately, we didn't have friendly fighter escort on this mission. We manned the jets, took off and headed to the entry point north of the target. We proceeded south as usual then made our final turn to the southwest for our attack heading.

---

[82]The bombs deployed fins to slow the bombs down after release minimising the risk of bomb fragmentation.

Everything was going according to plan as we dropped our bombs on the designated target.

Seconds after our bomb release, an F-5 fighter swooped down on us from above right. Mike put on a hard 6.5G turn into him hoping for an overshoot. That turn lasted about 10 seconds then 6.5Gs back to the left for about 10 seconds. The F-5 was much more agile than we were and nestled in behind us for the kill. Mike put on a great set of defensive maneuvers up to 6.5Gs, trying to get the bandit to lose his guns tracking solution. This was physically demanding on our bodies. The bandit didn't seem to want to let go of this kill as he kept stalking us. Our engines had been at maximum thrust ever since turning into the target. Our airspeed had varied as we bled off energy in the accelerated turns. We kept looking back over our shoulder to see the F-5 still behind and above us. He was compensating for our turns and narrowing on a final kill where his gun crosshair would linger on us. Mike kept the hard maneuvering up until we saw water approaching then relaxed the stick and proceeded wings level. We returned to NAS Cubi for the debrief. Apparently, our bombing was accurate and the fighter didn't comment.

23-28 January For the next five nights, we flew FCLPs in preparation for going back to sea. I flew with Barney, Waldo, Rocco and twice with Jeff.

Thursday 30 January Waldo and I walked aboard Midway just before she pulled out of port. I was the SDO for our first day at sea. I was confused with who does what, how to make changes to the flight schedule, who to call, who to keep informed, what to say, etc.… The aircraft carrier works with everyone knowing what is going on through communication. I experienced trial by fire that first day. You can read operational procedures in a book but learning it by doing it under pressure is a whole new ballgame. I was relieved to turn over the duty to the next officer.

Saturday 1 February I walked into the ready room to see the television turned on and broadcasting news from the United States. This was out of the norm. The tone of the ready room was somber as were the commentators on the TV. I asked the SDO for clarification. He said that the Challenger Space Shuttle had exploded on takeoff. That was shocking news. I watched the TV for a while soaking in the tragedy. It was impossible to comprehend. I didn't have time to consider it as I had two tanker flights with Waldo, one day and one night.

We were giving away gas to F-4 Phantom aircraft. They were great to watch as they rendezvoused on us for fuel. We had three arrested landings at the completion of our day flight. That helped Waldo get requalified and comfortable before our night mission. My first night tanker flight overhead the carrier was a real eye opener. There were aircraft everywhere. Some were separated by altitude, others merely by pilots avoiding each other with technique. You really needed to keep aware of what was happening around you. Waldo trapped on his first pass. It was good to be back down on the deck.

Sunday 2 February Barney and I flew a day tanking hop. I enjoyed flying with him after our time together in Cubi. We got two traps to help Barney get current on landings. Waldo and I had another night tanker flight. We gave away our gas then were given a two-day liberty pass to land back at NAS Cubi.

Wednesday 12 February Waldo and I were scheduled as the alert Alpha tanker, in a KA-6D. The airwing wasn't scheduled to fly but would if any Soviet or Chinese aircraft came near to us. In alert Alpha, the fighter crews needed to launch in 5 minutes, so they sat strapped in their aircraft on a catapult with the engine starter cart hooked up. They could be launched off the deck within five minutes given the word. We were required to be in full flight gear in our ready room, with our tanker airborne within fifteen minutes of notification.

A Soviet Bear D bomber[83] was inbound to pay us a visit and we were launched. We joined on an F-4 Phantom that was escorting the Bear. It was exciting to be flying so close to our Cold War nemesis. He was flying out of Vietnam and keeping track of us. The Bear carried missiles that could severely damage the USS Midway, if not sink it.

Thursday 20 February USS Midway was anchored off the coast of Pattaya Beach, Thailand. The squadron officers pooled money and rented

---

[83] The Bear bomber is a 50s design: https://www.bbc.com/future/article/20150225-the-worlds-noisiest-spyplane

a suite in a nice hotel to serve as the center of social activities ashore - it was called the Eagle Admin.

Thursday 27 February We had an AOM at 10am aboard the Midway. Everyone enjoyed great liberty in Pattaya. I spent the day preparing charts for flights in South Korea.

The USS Midway and her airwing spent the month of March 1986 off the west coast of South Korea. We were there in support of the Team Spirit annual exercises that demonstrated the strength of the forces that would defend South Korea if attacked. The exercises integrated the US and South Korean defenses from the Navy, Army, Marine Corps and Air Force. The perpetual defense of the South Korean peninsula was one of the most extensive and expensive military efforts in the world at the time. Our airwing worked with the air and ground forces in exercises that simulated events that could occur during an invasion from the north. Midway was scheduled into drydock for modifications at the end of March 1986. This was last time to get quality carrier operations completed before the repairs finished in December.

Sunday 9 March Pilot, K9 and I were on the Alert Alpha. He was a very intelligent, rather quiet pilot. K9 had graduated from Massachusetts Institute of Technology (MIT) in electrical engineering. He worked for Boeing after graduation helping to design the 757 and 767 cockpit layouts. While working for Boeing, K9 decided to join the Navy to fly.

K9 and I launched to give fuel to F-4's chasing Soviet Bears to the north of Midway. We were in full flight gear in the ready room when the call came to launch our aircraft. It took us thirteen minutes from leaving the ready room to the cat shot. We gave away our fuel to the Phantoms and returned to Midway.

Monday 10 March Rocco and I almost died this night in a mid-air with another aircraft above USS Midway. He was a senior LCDR, our MO and a legend in the Intruder community. Rocco was soft-spoken with a quick wit, brilliant mind and manner that made you feel right at home in the cockpit. I was inexperienced in carrier aviation operations and was learning as I went along. We flew day and night KA-6D tanker hops

76

together this day. Our tasking was to provide inflight refueling to aircraft in need. We gave away 12,000 pounds of fuel during a typical inflight tanking evolution.

Rocco and I served as the last night, overhead, recovery tanker. The recovery tanker flew in a counterclockwise, racetrack pattern on the port side of the carrier at 3000 feet MSL and within 5nmi. We were watching an F-4 Phantom that was having a hard time getting a trap. He kept boltering requiring another fifteen minutes of flight time to set up a long straight in approach before attempting another landing.

As the F-4 continued to bolter, he was getting low on fuel. We were told to position ourselves in front of the aircraft should he bolter again. Rocco was doing a great job of managing our position in the sky to help this F-4 crew but the weather started to deteriorate. We had been dodging clouds before but needed to descend to keep sight of the carrier. Rocco reset the radar altimeter to 1800 feet. The clouds cleared as we descended through 2300 feet MSL. We slowed our descent, leveled off and were in a left bank while focused on the inbound F-4 and Midway.

Suddenly, everything changed. I casually glanced over my right shoulder to ensure we were clear of traffic. To my great shock, there was an aircraft flying right at us with only seconds to go before impact. I yelled, "Right, right, right!" to urgently get Rocco's attention hoping that he would change a rapidly developing disaster. There was a split second as Rocco looked, then quickly pulled back on the stick to avoid the other aircraft generating a 3.6G accelerated climb into buffet[84]. Our hearts were racing, our breathing rapid as the Intruder climbed higher into the clouds and the clear night sky.

We exchanged glances of shock and relief while intently listening to the radio for a discussion of what just happened. There was silence. Certainly, someone must have just seen the near midair. The F-4 we had been hawking, successfully trapped. The remainder of the airwing aircraft trapped with us landing last with an OK two-wire. It was good to be back on deck. We made our way off the dark flight deck and down to maintenance control to take off our flight gear and discuss the events.

---

[84] As aircraft approach the angle of attack that stalls the wing, they exhibit rough flying qualities called buffet.

Rocco told the maintenance control chief that we had a near midair, apologized but told him he had overstressed the aircraft. The chief sensed the gravity of the situation and almost joyfully said, "That's OK. Don't worry about it. It's good to have you here. We'll do the inspections".

Rocco and I made our way down to our ready room to talk to the LSOs about the landing but more importantly to discuss the near mid-air with the Carrier Air Traffic Control Center (CATCC) representative. A CATCC controller made his way through the ready rooms after CAT III approaches to seek feedback and clarify events. Rocco asked the controller about the identity of the aircraft that nearly ran into us. The controller was not aware of the incident. We were shocked and explained the sequence of events and asked the controller to help clarify things with research with his staff. We later found out that there was an A-7E aircraft on downwind at that time but neither the pilot nor the approach controllers noticed the near-midair. We went to the aft wardroom and had ice cream to celebrate being alive.

Tuesday 11 March Waldo and I flew another tanker mission in the morning then the carrier pulled into port at Pusan, South Korea. Travelling by ship is an unusual experience. You leave one set of coastal scenery and after a week of travel in an empty ocean, another set of scenery appears. The first thing we noticed about South Korea was that barb wire was everywhere. South Koreans were serious about war. We took a thirty-minute bus ride to downtown Pusan where Eagle aircrew met in a bar and sang Beach Boys songs at the top of our lungs.

Sunday 16 March Waldo and I had a morning KA-6D tanker flight. We briefed for a night bombing mission in the afternoon. After takeoff, we preceded to a safe area, dropped two smokes in the water and dropped MK76 bombs on them. We climbed up to 15,000 feet and flew outbound from the ship. Strike called us with a Surface Surveillance and Control (SSC) mission[85]. We found a great number of tankers and freighters. I used the aircraft's FLIR to identify the ships. We found a Soviet Kashin class destroyer[86]. We descended to 2000 feet to get a better look.

---

[85] SSC missions assist the carrier in keeping track of all seaborne traffic in their vicinity.
[86] We often saw various Soviet ships in WESTPAC;
https://en.wikipedia.org/wiki/Kashin-class_destroyer

We started to head back to the USS Midway. I was able to get scenery shots of the South Korean coast on the FLIR. As we got closer to the carrier, there was a lot of discussion regarding an emergency beacon coming from a boat. Our recovery helicopter said that it looked like a fishing boat but it was too dark for him to tell if they were having any problems. Waldo and I were asked to make a FLIR reconnaissance overflight. We got a vector, saw it on radar, then on FLIR. Everything looked OK and we reported this to Strike.

Monday 17 March I stood the SDO duty for the first day of Team Spirit[87]. It started at 130am for the first brief. SDO for Team Spirit was one of the hardest duties we had. There were so many conflicts, problems and issues to resolve. One of our BNs was sick, another was on leave. We had many flights scheduled. More got added to the flight schedule. A few got cancelled. I was relieved of the duty at 9am by pilot Fig and went to sleep.

Wednesday 19 March Rocco and I flew a coordinated War At Sea Exercise (WASEX) with A7 and F15 aircraft. A KC-10 provided inflight refueling for the strike package. After refueling, we joined an A3 tactical reconnaissance aircraft. We were inbound to multiple ships and decided to overfly them. The weather was terrible. We descended to 200 feet above the water. Each of our aircraft took a separate ship to target. My target was all locked up with our radar and FLIR. Rocco did not want to overfly the vessel at low altitude. The ship triggered our onboard electronic warfare displays. We looked down to see a Soviet Kresta cruiser[88]. That was exciting to see. We turned around and flew back to Midway.

The second flight of the day is etched in my memory as it was so unusual; 130am brief time, 330am launch time for a night tanker mission with Rocco. We gave away fuel to an F-4 and headed back overhead the ship at 20,000 feet. We saw Halley's Comet about 30 degrees above the horizon in the western sky. We checked in with Strike with our fuel state.

---

[87] Team Spirit was a multi-national, multi-service military exercise to practice defensive tactics for South Korea.
[88] The Soviets were keeping an eye on us; https://en.wikipedia.org/wiki/Kresta_II-class_cruiser

Strike copied the information then called back a minute later, "Eagle 516, we are considering an option to shut down flight operations for the night. Do you have enough fuel to stay airborne and trap at first light in the morning?" Rocco and I were stunned. I replied, "Standby".

We needed to look at the fuel onboard, our fuel usage considering a climb to higher altitude, maximum endurance fuel flow at high altitude and time. After a minute of analysis, the numbers looked good. We estimated that we would trap with about 5klbs of fuel. "Strike, Eagle 516 can stay airborne until first light" was my reply. "Copy, 516, standby" was their response. After a couple of minutes, Strike came back, "Eagle 516, it's a deal. We will catch you at first light. Have a nice night". "Eagle 516 copies, we are climbing overhead Midway to flight level 410" was my reply. Strike confirmed, "Copy 516, FL410."

We climbed to FL410 to reduce our fuel consumption. The high altitude also provided a clearer sky to view Halley's Comet. At FL410, we were above any sort of atmospheric pollution or humidity. The heavens were clear, but Halley's tail was a blur of light. There was a ¾ full moon shining making our viewing more difficult. We darkened our cockpit lights to eliminate reflections on the canopy. We turned up the instrument lights every 10 minutes to check engine performance and fuel onboard. We felt extremely blessed to have one of the best vantage points in the world that night for viewing the comet. Hours passed as we flew in a racetrack pattern above the Midway.

At 5am, Strike called and told us to expect a 6am visual Case I recovery as it became light. Rocco pulled power to idle and we started a one-hour descent using practically no fuel at all. There were no other aircraft flying so we ruled the skies. We entered the downwind around 6am, conducted our landing checklist and started our turn at the 180. We rolled in the groove with a centered ball calling, "Eagle 516, Intruder ball, 4.9". "Roger ball, Intruder, welcome home" came from the LSO. We trapped, cleared the landing area and parked the jet.

When we got out of the jet, the flight deck was quiet. The carrier was just waking up, but no aircraft engines were yet turning. We went downstairs to maintenance control, completed the paperwork, took off our flight gear and headed to the aft wardroom for breakfast. That was an amazing experience!

Saturday 22 March Waldo and I briefed an unusual mission with a 10pm takeoff time for a 1030pm target time at Daegu (spelled Taegu in 1986). We were to coordinate with USMC units with LASER designators to conduct practice attacks in mountainous terrain. Midway was off the west coast of South Korea and Daegu was on the eastern side of the peninsula. We launched off the bow into complete darkness and started our climb to traverse South Korea at FL290. It took about 25 minutes to cross over to Daegu. We requested and received permission to descend to 15,000ft to work with the Marine controllers. We called the Marines on our back radio to start coordinating our practice attacks.

My initial radio call went unanswered, so I called again. After a while, a response was heard from the ground. We identified ourselves and our intent to work with the Close Air Support (CAS) team. A Marine responded, "There is no one available to work with you. Everyone is asleep. Sorry". I responded, "Ok, Eagle 510, departing to the West". We requested and received a clearance to climb back up to FL310 and return to Midway. A nearly full moon was behind us, lighting up a pretty, yet frigid, landscape below.

As we pondered the failed mission, we agreed that if we were in the mountains around Daegu, we would want to be in a warm sleeping bag. We checked in with Strike with our fuel state explaining that the mission was cancelled. We were sent to the Marshall stack to await our turn in the landing sequence. We commenced our descent on time, flew the approach and trapped aboard just before midnight.

Monday 24 March Pru and I were leading a two-plane back to Atsugi with pilot, Mark and BN, Toad[89] in the other aircraft. We briefed the flight and walked to the jet. A7E Corsair and F4 Phantom aircraft were bagging their last few traps on Midway as we manned our jet. We launched, joined with our wingman and enjoyed the beautiful night flight over the Japanese islands before landing in Atsugi.

---

[89] Toad and I were later classmates at USNTPS starting in July 1989.

When the squadron returned to Atsugi at the end of March 1986, families reunited and looked forward to months without carrier operations. While we wouldn't be going to sea aboard Midway, we had several detachments planned in the months ahead to keep us proficient in tactical flying.

Monday 31 March A group of us went down to the USS Midway to pack out our gear from our ten-man, bunkroom eight. As Midway was going into the yards for repair, everything had to be off the ship. The rooms had to be left bare. In our bunkroom, there was stuff that had been there for decades!

Wednesday 2 April K9 and I flew a visual low level in a tanker. I made up the route to fly through the rugged Alps of Japan. K9 was maneuvering up to the 3G limit in the heavy tanker. We were clearing ridgelines, rolling inverted and pulling down into the lower terrain. Low clouds began to make navigation difficult. We continued VFR on top of the clouds to Misawa.

We shot a PAR approach at Misawa and broke out of the clouds at 400 feet with 1/2 of a nautical mile visibility! That was an exciting approach! We did a waveoff, climbed up to FL 220 and headed back to Atsugi VFR on a southerly heading[90]. We shot a high TACAN approach to Yokota AFB, raised the gear and flaps on the missed approach and preceded to the break at Atsugi. We did a touch and go then a no-flap, no-slat landing to a full stop. That was a busy and productive training mission.

Thursday 17 April We were trained in the deployment of Harpoon missiles. There was a great deal of information shared. Our AQ branch enlisted men also received Harpoon training and directions how they needed to prepare our aircraft to shoot the missile.

Thursday 24 April Barney and I flew a SSC flight south of Atsugi in Sagami Bay. We conducted practice AGM-123 Skipper II missile[91]

---

[90] International flight rules allow VFR flight above FL180, but it isn't permitted in the USA.
[91] The Skipper was a 1000lb Laser Guided Bomb (LGB) with a rocket motor on the back; https://en.wikipedia.org/wiki/AGM-123_Skipper_II

attack runs on ships. A typical Skipper run would start from 15 nautical miles at an oblique angle to the ship's broadside. We selected the weapon for a practice attack and stepped the mission computer into attack, even though we were not carrying the weapon. I acquired the ship on radar, completed a FLIR handoff to designate then tracked the ship with the FLIR. My target tracking provided Barney with steering and release symbology on his displays. The simulated weapon release occurred several nmi away from the ship at which time Barney put on a 4-6G turn away from the ship. My task was to track the ship in the turn until simulated impact. The Skipper missile was designed to track our LASER spot, but we didn't use the LASER on the ships to protect their crews.

As we headed toward Oshima Island for mining practice, the aircraft had a combined hydraulic system failure. We had a flight hydraulic system that would power most requirements but needed to land As Soon As Possible (ASAP). We pulled out our NATOPS pocket checklist and went through the emergency steps. I contacted Yokota approach and carefully explained, "Yokota, Eagle 501 is declaring an Emergency. We need a PAR into Atsugi, we will take an arrested landing, shut down our engines and need to be towed from the runway". We had a successful arrested landing and were towed off the runway.

Monday 5 May Waldo and I flew a warm-up hop as he had been on leave and hadn't flown in over a month. We were scheduled to perform missile attack profiles with a US Navy ship south of Atsugi in Sagami Bay. We tried to contact the ship by radio but were unable to reach them. We penetrated the clouds over water near the ship and broke out in the clear at 1000 feet MSL. We continued down to 200 feet flying around the ship in a hope to gain their attention and establish radio contact. We circled around the ship for ten minutes, but the ship failed to contact us. We flew back to Atsugi and executed a pair of PAR approaches.

Wednesday 7 May I was told in the morning that I was flying with K9 to South Korea the next day for a day/night SRTC low level. I briefed at noon for a flight with Tom, my first in Japan with him. We flew TC in the mountains on the new Black route enjoying the beautiful scenery along the way. We flew south to the ocean and up the beach at

200 feet AGL. After the flight and dinner, I stayed up planning for South Korea until 130am.

Thursday 8 May I woke up at 5am to complete our flight plan and weather brief. I made a copy of our mission information for our wingman BN, Jim, then conducted our brief. We were running late to takeoff for a hard entry time on the IR 982 low level in South Korea. Our aircraft had Master Caution problems and we took off 15 minutes late. Our engine throttles were set for maximum continuous power, EGT 594 degrees C. We entered the IR982 on time and flew TC as single aircraft. We conducted no-drop bombing using the RBS system like the Spokane training at VA-128. We proceeded to Osan AFB[92] and landed. We were given a ride to the BOQ, had lunch at the Officers Club then prepared for our night low level. The night flight went well except we could not get into the Koon-Ni bombing range. We conducted two, section PAR approaches instead.

Friday 9 May We woke up at 7am, had breakfast then raced over to the aircraft for our brief, taxi and take off. We had four runs at the RBS bombing range and flew back to Atsugi. We had an incredible vision of Mount Fuji as we flew by at 16,000 feet on a clear blue day.

Wednesday 21 May HMAC and I led a section of Intruders from Atsugi to NAF Misawa, 1.2 hours north for a bombing detachment. Our wingman crew was Pru at the flight controls with Six-Gun[93] as his BN. We flew into Misawa on the 21st and flew out on the 23rd of May. The detachment was HMAC's idea to improve our bombing currency without the huge logistics of a major detachment with many maintenance personnel. All we needed was fuel (provided by NAF Misawa), 25lb MK76 practice bombs (supplied by our Ordnance team at Misawa) and routine maintenance (provided by our small group of maintenance personnel). Our maintenance team was pre-positioned ahead of our arrival via a C-9 Navy transport aircraft.

As carrying bombs out of Atsugi was prohibited due to the great population surrounding the base, we needed to go somewhere else to

---

[92] Osan AFB was located 50 miles south of the Korean Demilitarized Zone (DMZ).
[93] Six-Gun was a LCDR that liked flying with his own personal firearm when needed, thus the callsign.

maintain bombing proficiency. As events during a bombing run require split second placement of cursors and crew coordination to ensure accurate bomb placement, routine practice was essential for accuracy as well as safety. Night bombing, using our FLIR system, required yet more practice and coordination. Our squadron had gone to the Philippines and South Korea in the past as they didn't have the restrictions that existed from Atsugi. Misawa was a great base for bombing as it was positioned adjacent to the Pacific Ocean. We took off and were quickly over water. We could return over water in the event a bomb wouldn't release as intended, referred to as a 'hung' bomb.

On the takeoff from Atsugi, Six-Gun recommended a section takeoff[94]. That was a brilliant suggestion as we entered the clouds at 1000 feet MSL and remained in heavy clouds until 14,000 feet MSL. That gave Pru a workout as he needed to fly in tight formation so that he didn't lose sight of us. The clouds were bouncing us around making close formation difficult. Once we were above the clouds, we all relaxed and had an enjoyable flight. As we arrived near Misawa, we made a familiarization pass at the Ripsaw bombing range before Rocco and his flight arrived for their bombing practice. Ripsaw was ten nautical miles north of Misawa with the targets just inland from the beach. The threat of any problems with a practice bomb was minimal. A bomb that landed long would fall into the Pacific Ocean. A bomb could land a mile short and still be on Ripsaw Range land.

We came into the break at Misawa then had a flap/slat deployment problem on downwind that HMAC was able to fix. We went to the O' Club for lunch then time off before briefing for our evening flight. The BOQ rooms were nearly identical to the Atsugi rooms. We flew to Ripsaw Range and dropped twelve MK76 bombs.

Thursday 22 May We had an early morning brief then breakfast. We were all eager to do more bombing at Ripsaw. HMAC and I completed pop-up, shallow dive releases with his gunsight designation.

---

[94] A section takeoff is when two aircraft take off simultaneously. The lead aircraft assumes one side of the runway while the wingman stays on the other. The wingman adjusts his power to remain in a close parade position relative to the lead aircraft. After raising the landing gear, flaps and slats after takeoff, the wingman may increase his aircraft separation or maintain a close position in the event of poor weather.

He was a bullseye bomber and loved to pull hard off target. We had lunch then time off. We had a night, two-plane flight at Ripsaw to drop twelve more MK-76 bombs. Our aircraft was the lead for the formation. We broke up into single aircraft enroute to the target area.

We had a bullseye bomb on our first release at 500 feet, 420 knots groundspeed. That felt great! We had five more low altitude drops with good results. We tried to work the high, dive-bomb pattern and ran into problems[95]. The weather was making target acquisition difficult with poor visibility and clouds floating around the target area. We were rolling in from 12,000 feet MSL to achieve 40° dives. Our first dive was 35° passing 10,500ft. HMAC committed for the bomb release and began his pullup at 5,000 feet MSL. We didn't release a bomb due to a 114-computer error code, Excessive Noise. HMAC was not happy as he felt his flying and reticle placement were smooth.

Our next dive started at 37° at 10,500 feet with a pullup at 5500 feet MSL. Our hit was 200 feet short of the target. HMAC was not impressed. HMAC was dodging clouds to maintain sight of the target. We rolled into a 30° dive from 9500 feet MSL but HMAC aborted the run due to clouds and we proceeded back to the low pattern. He wanted to try pop-up, boresight deliveries[96].

Our first run was inbound at 5nmi and cleared hot. HMAC maneuvered the aircraft into an offset to the right then back left into a 15° dive from 3000 feet MSL. The pullup started at 1000 feet MSL but we got another 114-computer code. HMAC had enough of that nonsense and switched to a manual gunsight reticle for deliveries where he would manually pickle the weapons to ensure they came off. We dropped several bombs then joined our wingman to discover that each of us had a hung bomb. We completed solo, precautionary, straight-in landings back at Misawa. We made it over to the Officers Club for dinner.

Friday 23 May Our first morning bomb run was another bullseye. HMAC and I were ecstatic! He wanted to do more pop-up, shallow dive work but the bombs were dropping short as shown on my FLIR. We had another bomb impact at 23 feet with a release of 200 feet above ground

---

[95] I recorded part of the cockpit audio on a cassette tape providing specific flight details.
[96] Boresight deliveries were pilot designated but computer commanded weapon releases.

and 473 knots ground speed. We did a pop-up boresight to BN hand off to my FLIR reticle for the last bomb and got another bullseye. We improved our bombing accuracy in a couple of days.

The scoring tower asked for a flyby so they could take pictures. We agreed to their request and started our inbound run from eight nautical miles. HMAC led the two-plane in at 200 feet above the ground and at 490 knots indicated. The tower loved it saying that the F-16's had nothing on us. We flew back in Atsugi without refueling as all the bombs released as designed. If we had a hung bomb, we would have landed at Misawa and had it removed prior to our return to Atsugi.

We pulled the tape from the aircraft Video Tape Recorder (VTR) after landing in Atsugi. It had great footage including the last pilot to BN FLIR handoff for a bullseye. Our AQ sailors loved it. Their hard work had contributed to the bullseyes. The aircrew went to the O' Club for dinner to celebrate.

Monday 23 June Our squadron had a brief from the Medium Attack Weapons School (MAWS) at training. MAWS also conducted training with the Aviation and Armament divisions of maintenance.

Monday 30 June We had a squadron AOM then the XO and I flew the Brown low level. MAWS gave us an ACM brief with the best strategies for the Intruder pilot to use when faced with an attacking enemy fighter.

In mid-July, we had a detachment to Osan AFB and Suwon AFB[97], South Korea. I was transported by a Navy C-9 aircraft into Osan AFB and checked into the BOQ. Our Intruders landed and were maintained at Suwon AFB. We took a bus there each day to fly our missions. Suwon was the proud home of an Air Force A-10 Warthog squadron that aided with our detachment. The A-10 guys couldn't have been more helpful. In fact, we enjoyed their company so much that we didn't want to leave. It is important to note that A-10 pilots are a very special breed[98]. They were tasked with stopping a North Korean advance

---

[97] Suwon AFB is located 20 nmi south of Seoul and 36nmi south of the DMZ.
[98] Watch this TED talk to help think like an A-10 pilot;
https://www.youtube.com/watch?v=5WtQqKrbmKc

in the event of war. The A-10 aircraft is designed to kill tanks and other armored vehicles with an amazing 30mm cannon built in the nose. When the pilot placed his Heads-Up Display (HUD) aiming reticle on a target and pulled the trigger, it was destroyed. The A-10 was a low and slow airborne jet fighter that supported the troops on the ground with real-time firepower. The pilot was surrounded by a bulletproof capsule to protect him from enemy ground fire.

Wednesday 16 July We had an area brief for flights out of Suwon. On our first arrival to the main gate at Suwon, we were surprised to see a manned tank inside the gate with its barrel pointed directly at us. Had an enemy tried to come through that gate, they would be met by machine gun fire then an artillery shell. That would probably stop them from coming on base. We were impressed if not taken back by their no-nonsense approach to base security.

Thursday 17 July The XO and I had a 730am brief for a mission to drop live MK82 bombs on the small target island of Chik-Do off the west coast of South Korea. We scored a bullseye on the island with a high dive. It was good to deliver live bombs again as that was what we were trained to do. It took practice for the ordnance team and the aircrew to deliver bombs on target The XO and I flew a night IR982 low level finishing at Kooni target range for a MK76 bomb release.

Friday 18 July The XO had an 11am brief to fly the IR982 again followed by the inert bombing pattern at Kooni range. We landed and had enough time for a hot dog before launching again for more bombing at Kooni range. The second takeoff was in bad weather, but it cleared as we made our way to Kooni. We had two bullseye hits.

After flying, our squadron aircrew and the A-10 pilots played Crud[99] in their ready room with very lively competition. They asked and we agreed to play 'Combat Crud' which meant that physical interference between players was permitted. As everyone was having so much fun, the A-10 pilots suggested we travel with them down to Osan to visit their

---

[99] Crud is played with a pool table, a white cue ball, a striped ball and two opposing teams with two players at a time trying to sink or prevent sinking, the striped ball. The rules can be found here; https://en.wikipedia.org/wiki/Crud_(game).

favorite club outside the Air Force Base. The JOs agreed and piled into a group of taxis.

The club that the A-10 guys led us to was off the main street, empty and quiet when we walked in. That changed as soon as the owner saw the A-10 pilots walk through the door. Videos of Warthogs blowing up targets came on all the video screens with accompanying loud music. The beer started flowing and the dancing began. The bar came to life! We loved these guys. As the videos were playing, the A-10 pilots got really excited by things exploding. It was great to see their enthusiasm. Their job would be one of the most difficult in a war with North Korea. They needed all that passion to stop the aggression.

Monday 28 July I delivered an AOM presentation explained how we would change the way we shipped and handled our aircraft's search radar, IMU and DRS. Better maintenance procedures and diagnosis of problems ultimately led to improved weapon system performance. The aircrew welcomed the changes.

We start using specially designed support equipment to transport the search radars, IMUs and DRSs. We acquired our own feedhorn alignment tool to fine tune the radars along with establishing a radar warmup radar area to confirm weapon system accuracy before takeoff. This was common practice back at Whidbey Island but the high tempo of flight operations from Midway resulted in this type of maintenance taking back seat to ensuring the aircraft could pass fuel to fighters.

Saturday 9 August A group of Eagles climbed up Mount Fuji. We signed up for a Morale Welfare and Recreation (MWR) tour that took us by bus from Atsugi to the base of Fuji. We climbed during the day, had an overnight sleep at the summit and the bus picked us up the next day for our return journey to Atsugi.

The summit of Fuji is at 12,388 feet above sea level so the climb required us to pace ourselves. As we climbed over 10,000 feet in elevation, hypoxia started to set in. The climb was a good test of fitness. Vern was a very fit runner and finished with the very impressive time of 2 hours, 30 minutes!

Wednesday 20 August Fig and I flew an Intruder down to Kadena AFB for a short detachment. Okinawa was a beautiful place and the flying was scenic with surrounding coral reefs and aqua blue water. Kadena had two 12,000 foot long, parallel runways with plenty of parking and hospitable Air Force staff.

Thursday 21 August Fig and I completed a day training flight from Kadena with other squadron aircrew. While gifted in many ways, one of his unique skills was close formation flying. Fig tucked our aircraft in very close to the lead aircraft. At times, it felt like our aircraft wingtip was only a few feet from touching the lead despite our great speeds. He flew much closer formation than any other pilot in my experience. It looked 'Blue Angel close' to me. Fig's acute responsiveness on the stick and throttle along with his quick mind allowed us to remain safe despite the risks.

Our squadron traveled from Atsugi to NAS Cubi in late August for a detachment. It was always great to be in the Philippines (PI), but it has never rained harder in my experience. September is in the middle of the rainy season in the PI. It was like someone pouring buckets of water on your head, continuously. It was crazy. We learned to stay indoors as much as possible but flying required us to get wet. We walked to our jet in full flight gear and were drenched before we closed the cockpit canopy. Most of our flights during this time were FCLPs so we'd stay in the pattern with heavy rain if the visibility allowed it. With an LSO controlling us, normally we flew with a mile of visibility. On occasion, it might go down to ½ of a mile, but we would maintain contact with the airfield environment and keep flying. I was crewed with KO during this detachment for ten flights. We completed day and night attack missions dropping MK76 bombs but the rest were in the FCLP pattern, mostly at night.

Friday 19 September Beav and I flew to Osan AFB, South Korea for a cross country. As it was a training flight, we had waypoints designated for us to collect radar and FLIR imagery with our onboard VTR system. I selected the waypoints from 1:50,000 scale maps of cities along the high-altitude airway route of flight. The imagery was used to create training folders for the other squadron aircrew to help maintain their target acquisition proficiency.

Tuesday 23 September KO and I completed a 3-hour flight from Atsugi. The mission was a WASEX near NAF Misawa at the northern end of Honshu with our return to Atsugi. Our next destination was a return to NAS Whidbey Island and NAS Fallon, NV.

## Chapter 6

## NAS Fallon[100], F/A-18 Hornets, USS Kitty Hawk, Australia

*A Combat Air Patrol (CAP) had been in place for eight hours without any activity at all. Suddenly things got very exciting as the E-2C picked up a low altitude contact moving at supersonic speed from the EAST toward Midway! The CAP was ordered to reposition to the east to counter the threat. Unfortunately, there was no possible way for the Hornets to reach the inbound threat in time. The USS Midway had lost the exercise as the Australians outsmarted us.*

October 1986 Personnel from the A-6E, EA-6B Prowler[101] and E-2C Hawkeye[102] squadrons flew on military transports from Japan to the USA. The Japan-based aircrew needed to train with the new F/A-18A Hornet fighter-attack squadrons that were joining Airwing Five; the VFA151 Vigilantes, the VFA192 Golden Dragons and the VFA195 Dambusters[103]. The Hornets replaced the aging A-7E Corsair and F-4S Phantom II aircraft. The Intruder and Prowler aircrew initially went to Whidbey Island to fly simulators, receive briefings on the latest tactical procedures and low altitude flight safety awareness.

The airwing detachment site was at the Navy's new tactical aircrew training facility, Strike University[104], at the Naval Strike Warfare Center, NAS Fallon, NV. The goal of the detachment was to refine airwing tactics and strategies while incorporating the Hornet aircraft and aircrew. The E-2C's crews with their long-range radar dish provided situational awareness of simulated enemy fighters and directed our fighters to engage them. The Prowler crews electronically jammed the

---

[100] NAS Fallon trains all Naval tactical aircrew;
https://cnrsw.cnic.navy.mil/Installations/NAS-Fallon/
[101] The electronic warfare Prowler;
https://en.wikipedia.org/wiki/Northrop_Grumman_EA-6B_Prowler
[102] The E-2C had a long-range, airborne radar;
https://en.wikipedia.org/wiki/Northrop_Grumman_E-2_Hawkeye
[103] The Dambusters name came from the WWII bombing raids;
https://www.iwm.org.uk/history/the-incredible-story-of-the-dambusters-raid
[104] Strike 'U' provided the best training;
https://www.intruderassociation.org/strikeu.html

enemy defense radars and simulated firing High Speed Anti-Radiation Missiles (HARM)[105] to destroy them.

Tuesday 21 October Our squadron was loaned Intruder aircraft from VA-95 at NAS Whidbey Island while flying at Fallon. The JOs wondered if they loaned us their hangar queens[106] that could barely get airborne. We were pleasantly surprised that they loaned us their best aircraft which was much appreciated.

KO and I had a mid-morning brief to shoot 5-inch Zuni Rockets at a nearby bombing range to regain our currency. KO aggressively maneuvered the aircraft during the flight. As we were inbound and outbound on our bombing runs, KO was making 5G pulls and jinking maneuvers at low altitude as if dodging Anti-Aircraft Artillery (AAA) or SAMs. We continued launching the rockets with shallow dives and pilot designated releases. We were demonstrating our combat readiness! The Zuni rockets were awesome and a real thrill to fire. We couldn't get training like that in WESTPAC.

Wednesday 22 October We commenced the airwing training with a bombing competition for the Intruder and Hornet squadrons. We were given a day flight to practice our bombing with MK76s at a nearby bombing range. The competition started the following day with graded day and night flights. The four closest bomb hits were entered into the competition. The four hits included a 40-degree dive bomb, a 200-foot AGL level laydown, a 15-30° accelerated loft[107] from a 200-foot AGL run-in and a night release of our choosing.

KO and I had an excellent aircraft with a functional FLIR and LASER, so we planned a pilot to BN FLIR handoff for the 40° dive. KO designated the target using his HUD aiming crosshair, I moved it closer to the target by slewing the FLIR crosshair and redesignating it. Once

---

[105] HARM permitted long range attack of enemy radars; https://en.wikipedia.org/wiki/AGM-88_HARM

[106] A hangar queen is an aircraft that has significant maintenance issues or is waiting on parts before it can fly again. Unfortunately, these aircraft are often cannibalized for parts to keep the other squadron aircraft flying.

[107] The loft release threw the bomb upwards into a parabolic flightpath thereby increasing the range of the weapon.

redesignated, I tracked the target center with the FLIR crosshairs. The tracking served to provide velocity updates to the mission computer prior to the bomb release point. The bomb came off and we got a very good hit. We descended into the low pattern to use the new Continuously Calculated Impact Point (CCIP) software mode of the mission computer. CCIP basically allowed the pilot to drive the gunsight reticle to the target with the weapon selected and the mission computer in attack. The mission computer calculated changes to the impact point as the dynamics of the flight profile changed. The pilot released the bomb when the reticle reached the target. KO scored a very accurate, level laydown and headed downwind to set up for a 20° loft release.

I acquired the target on radar, then transferred the designation to the FLIR with LASER tracking and velocity correcting. KO engaged the commit trigger on his stick and waited for the computer determined bomb release point. He pulled back on the control stick to initiate the climb to 20° nose-up. As we approached 20° and the release point, KO grew concerned that the bomb wasn't going to come off at the right moment. He pressed his bomb pickle button in the hopes of instantly releasing the bomb. We scored a bullseye on that release and were ecstatic. It is not clear if the bomb came off as commanded by the mission computer or by KO's pickle button. It was amazing that KO was so in tune with the aircraft and weapon delivery that he was confident to override the mission computer about a release point. That was KO. He was a genius that could do things like that[108]. Our day bombing had gone well but we needed a good night bombing result to be in the running for top honors. The aircrew rhetoric became livelier as the night missions approached.

The target for the night bomb deliveries could be illuminated for night weapon deliveries. The Hornet pilots wanted the target illuminated, the Intruder crews did not. A discussion about the merits of unlit targets in combat led the airwing staff to decide that the target would not be illuminated for the night bombing. As the target was unlit, KO could not designate it for me. I was able to find it on radar then hand off to our FLIR using our LASER for ranging. We delivered the bomb in a 30° dive and scored a hit of 23 feet.

---

[108] When KO graduated Navy Nuclear Power school years later while training to be the first CO of the aircraft carrier, USS George Bush, his final grade point average was the highest ever achieved at that incredibly demanding school.

We wondered how well the Hornets would manage delivery on the target. Not many Hornets had FLIR at that time and most crew were unfamiliar with using it. Several Hornets returned to Fallon without dropping bombs on the target. We felt comfortable with our bomb deliveries but knew that the Hornet had a superior weapon system design and were known to be bullseye bombers. The results of the competition were kept secret until an airwing party later in the week.

Once the bombing competition concluded, we got down to the real business of opposed, multi-aircraft strikes that simulated combat operations against preplanned targets. The targets were located inside the Fallon restricted airspace bombing ranges allowing release of various ordnance types. The strikes were opposed by the Strike U aggressor F-5 and A-4 aircraft. Each aircraft carried a Tactical Aircraft Combat Training System (TACTS) pod that provided real-time aircraft position and weapon release information. The aircraft positions throughout the strike were recorded to assist during the debriefs. These were held in restricted access auditoriums that contained large screens to display the entire Fallon restricted area airspace and the aircraft flight profiles. This enabled us to learn from our successes as well as our mistakes. We gained a bird's eye perspective of what we could expect from our new Hornet escorts on our missions. The Hornets were very agile and their engines didn't leave trails of smoke like our prior F-4 escorts. The Hornets appeared to maneuver well in sections and divisions and did a great job of protecting us from the aggressors. We flew several day, opposed strikes carrying a variety of inert bombs. The training was priceless.

Wednesday 29 October KO and I flew a day, opposed strike then prepared for our last flight. It was a night, nineteen aircraft strike to attack a simulated naval facility. The nineteen aircraft included a Command-and-Control E-2C, a pair of EA-6B Prowlers, Hornets serving as fighter cover, HARM shooters and deception elements and four Intruders to put bombs on target. This was not an unusual scenario. The airwing was often tasked with supporting Intruders as they served as the night bombing aircraft. Our intelligence scenario placed Cold War vintage SAMs in the B-17 target area so we chose a low altitude, night ingress to utilize terrain masking and darkness to limit the threat.

Simulated HARM launches were planned to destroy the SAM sites if they were radiating. The deception aircraft simulated an inbound strike away from the actual strike package to add confusion for the enemy. The EA-6B had good effectiveness jamming the enemy defenses.

The Intruders entered the B-17 target complex from the south at 500 feet AGL and 420KGS. KO and I were the fourth aircraft in a stream raid of four Intruders attacking the target with one minute weapon delivery intervals. The lead aircraft crew was the XO/Six Gun, the second aircraft was Tom/Bart and the third was K9/Bing. As always with SRTC, our cockpit dialogue was lively as we crossed over the ridge lines with elevations to 6908 feet MSL before descending toward the Fairview Valley floor at 5000 ft MSL at 12 nautical miles from the target. We were less than 2 minutes from the target yet the mountains were obstructing my radar view. I was anxious as the time to provide KO with updated steering to a bomb release point was rapidly disappearing. I needed to find the target, sweeten the radar designation then hand off to the FLIR and start an INS velocity correct. KO needed enough time to adjust the aircraft's flight path to reach the correct weapon release point.

As we got closer, the target appeared on radar then was obvious on the FLIR. I completed the handoff to the FLIR and had about ten seconds of tracking for the velocity correct before reaching our weapon release point. KO followed steering then snapped on an accelerated turn to the northwest after weapon release as I tracked the target on FLIR. The FLIR confirmed a good hit on the target. Once clear of the target area, we slowed down, climbed and contacted Fallon Tower for our return to landing. The debrief was a great conclusion to our training. The Intruder crews were feeling very satisfied with our performance on the night low level and weapon releases. It was rare if not impossible to fly SRTC at 500 feet at night in WESTPAC. The other airwing crews were also pleased with the events of the night.

The airwing had a party at the NAS Fallon Officers Club at the end of the detachment. The mood was jubilant as the training had been superb. We were enjoying the amazing Hornets and the protection provided by their pilots. The results of the bombing competition were announced with KO and I winning the event. Our whole squadron was excited as the old Intruder aircraft had demonstrated the accuracy of the airframe despite its age. Credit went to VA-95 for providing the excellent jets for the detachment. I imagine the Hornet crews were surprised by the results but the night unlit target took many of them out of the competition. It was not permissible to release a weapon on a target that couldn't be positively identified. Radar bombing was not something that the Hornet community had time to train for and they didn't have FLIRs on most aircraft.

Monday 10 November KO and I flew the newly designated Blue Route logging 2.6 hours. The route took us at a medium altitude from Atsugi to the entry point west of Mount Fuji. The route proceeded in a northerly direction through the Japanese Alps toward Niigata before exiting and returning to Atsugi through Yokota AFB airspace.

Wednesday 12 November KO and I flew an ACM section flight with K9 and Bing. Our two sets of competitive aircrews generated demanding flying[109]. We took off from Atsugi and headed 50 nmi south over the ocean to perform our maneuvering. After reaching the area we split up and traveled in opposite directions for a minute or two before heading inbound toward one another. The first aircraft to see the other could start his maneuvering to gain an advantage. We strained to see the other aircraft then flew directly at them to remove any advantage. We often passed within 500 feet of the other aircraft with closure speeds of 900+ KGS.

Once the pass occurred, it was critical yet at times painful to maintain sight of the other aircraft. We pulled 6.5Gs requiring our bodies to contort in all possible ways to keep that visual contact. Our aircraft made a few passes at one another with decreasing speed ultimately resulting in a 'rolling scissors'. As both pilots were exceptional, the fights often resulted in a draw. We reset our positions and started a new engagement. We completed six engagements before returning to Atsugi for fuel considerations. While these flights were great in lifting our spirits, they were also physically demanding and usually resulted in a good night's sleep.

---

[109] KO later served as the first CO of the aircraft carrier USS George H.W. Bush. K9 later graduated first in his class from U.S. Naval Test Pilot School. Bing retired after 30 years with tours to Japan, Saudi Arabia, Italy, Chile and Africa.

Thursday 13 November KO and I flew another ACM hop with Pru as the pilot of our wingman. The training we completed at Fallon paid off as the CO felt comfortable allowing us to fly ACM. In the past there was concern that ACM would result in a mishap as the crews hadn't been properly trained. Morale in the squadron started to soar after we were allowed to fly ACM.

Sunday 16 November We all had an early night to get ready for a detachment to NAF Misawa the next day. The detachment plan was for crews to complete night FCLPs in preparation for our return to flight operations aboard Midway in early December.

The F-16 Wing stationed at Misawa invited us over to their ready room for drinks and nibbles. We were honored to be invited. Upon arrival, we noticed a pool table in the room. That caught our interest as our squadron were experienced Crud players and had enjoyed the game with the A-10 pilots in Suwon. As the night went on, the F-16 pilots asked us if we would like to play Crud. We whole-heartedly agreed. They asked if we wanted to play Combat Crud allowing players to block their opponents and rank was thrown out the window. We agreed to that as well. The game started out with a bang. We were not ones to shy away from a fight and surprised the Air Force pilots with our enthusiasm and physical play. Long story short, we won the game of Crud and the F-16 pilots were upset that Navy pilots beat them on their home court. We were pumped as we walked away celebrating our success. We ended up at the Officers Club, did 'carrier quals'[110] on the tables then wrestled and finishing the night up with dancing. It seems we had a lot of energy and needed to get rid of it somehow. I suspect the other patrons of the Club wondered who the crazy Navy guys were.

Saturday 22 November KO and I had our last night bounce feeling confident to return to the USS Midway for actual carrier operations.

Friday 5 December The USS Midway had been in the shipbuilding yards of Yokosuka since April undergoing modifications.

---

[110] Carrier quals involves aviators running and jumping onto a table on their stomach, while trying to catch a towel or other simulated cable with their feet before sliding off the table.

The most dramatic modification was the addition of large sections of steel called 'blisters' that were welded to the hull below the waterline to reduce the roll and pitch of the flight deck[111]. Midway sailed south of Atsugi yet was close enough to enable efficient requalification of our airwing crews.

KO and I flew out to Midway, bagged two traps, two catapults and returned to Atsugi with a flight time of 1.1 hours. The modifications to the Midway did not seem to have a noticeable impact on the way the ship felt on approach or landing. Later that night, KO and I had a 5.3-hour flight that resulted in 2 catapults and three traps with us staying onboard. When I saw this flight in my logbook, it didn't make sense. Why would we have 5.3 hours of flight time to get 3 traps? I couldn't recall the events of the flight, so I emailed KO. He responded after 34 years with the following words affirming his still amazing mind;

"We headed out to the ship as (high) overhead tanker for the air wing CQ. I remember hanging on the blades at 30K ft for nearly 4 hours before getting the call to "Charlie" to be in the pattern in 15 minutes. We then went to idle and took about 15 minutes to descend to the case 3 approach. Then we got 3 traps and 2 cats and called it a night."

Tuesday 16 December KO and I bagged four daytime catapults and traps and waited for night to set in. The night was memorable for several reasons, the first being a 100% full moon on a clear night. In carrier aviation, it was called a 'Commanders Moon', as it was so bright that flying around the carrier was almost like flying in the day. 'Even a Commander would fly with a moon like that' was the joke in naval aviation. Some senior officers were accused of being shy of flying on dark nights.

KO and I were scheduled to get night traps. We hot seated into a turning jet, meaning we swapped seats with a previous crew that left the engines running. The BN in the turning jet got out first. I climbed into the jet and immediately plugged into the Intercom Communication System (ICS) to discuss the aircraft status with the off-going pilot. He

---

[111] Japanese naval engineers recommended against the 'blister' modification as it would magnify the rolling problems with the Midway in certain sea states. They were correct. Our engineers should have listened to them.

commented on problems the crew had noticed while they were airborne. Once I'd received the brief, the off-going pilot got out of the jet and KO climbed in, closed the canopy and plugged into the ICS. I briefed KO on the comments from the off-going pilot. We hot refueled the jet, went through our system checks and KO let the yellow shirts know we were ready to taxi.

We taxied out of our parking place and toward the catapult. Our takeoff checks were nominal. The Catapult Officer did his cross checks and signaled for our launch. We raced down the moonlit track, good cat shot and end-speed, the gear was coming up in the climb. I checked in with approach, "Approach, Eagle 507, airborne". "Roger Eagle 507, climb and maintain 1200 feet, when comfortable, turn downwind heading 180". I responded, "Eagle 507, climb and maintain 1200 feet, left turn 180".

The reflection of the moon on the water was beautiful and very bright. The carrier was easily seen. It wasn't as bright as day but the next best thing. We had kept our flaps down for the climb and pattern as we wouldn't be going very fast or very far. As we were abeam the stern of the ship, Approach called stating, "Eagle 507, expect a five-mile base turn". I replied, "Eagle 507, copy. Five-mile base turn". KO wasn't happy. "We could turn in here. This is easy" was his reply. KO was an exceptional pilot and probably could have turned in abeam the ship, just like in the day pattern, and completed a fine approach and landing. We decided to go along with approach on the base turn. We dropped our landing gear and completed our landing checklist as we neared five miles.

At five miles, approach called as promised, "Eagle 507, left turn to 090". "Eagle 507, 090" was my reply. KO turned on base commenting that this was too far out and wasting time. approach called, "Eagle 507, left turn to final bearing 347[112]". I echoed the call. We rolled out of the turn on the final bearing of 347° at 5 miles and held our altitude until 3 miles. We flew the pass on instruments despite frequent looks to the brightly lit carrier and meatball showing us on glidepath. "Eagle 507, ¾ of a mile, call the ball" came from the LSO. "Eagle 507, Intruder ball, state 6.0" was my reply. "Roger Eagle, 25 knots down the angle" replied

---

[112] The USS Midway flight deck is angled 13.5° left of the ship's bow and course through the water.

the LSO. KO flew a great pass and we trapped. The deck was bright, much brighter than on other nights.

We got out of the wires and headed back to the Cat. We verified our weight, went through the Before Takeoff checklist and taxied into the catapult shuttle. On signal, power to military, instrument and Warning/Caution light check then the anti-collision light came on. Down the cat we raced and once again were airborne into the moonlight. With our airspeed fine, gear up, climbing, I repeated our coordination with approach. All had proceeded well and we hoped to secure quite a few traps at that rate.

When we reached the abeam position, approach called us, "Eagle 507, would a three-mile turn to base be sufficient for you"? KO got excited and gave me a thumbs up. "Approach, Eagle 507, a three-mile base turn would be fine". "Eagle 507, approach, expect a three-mile base turn". KO dropped the gear and we went through the pre-landing checklist. As advertised, at three miles approach called, "Eagle 507, turn left 090". "Eagle 507, left to 090" was my reply. On the base turn, we looked much closer to the carrier than our last pass. As mentioned before, KO was an exceptional pilot and we didn't anticipate a problem. Approach called, "Eagle 507, left turn to the Final Bearing 347". I repeated, "Eagle 507, left turn to the Final Bearing 347". KO completed the turn and started our descent at 700 FPM. While tempted to fly a visual approach, we continued to fly an instrument approach with frequent, outside peeks of the carrier. The LSO called, "Eagle 507, ¾ of a mile, call the ball". "Eagle 507, Intruder ball, state 5.4" was my reply. The LSO replied, "Roger Eagle, 25 knots down the angle". KO flew another great pass and we trapped as before.

After clearing the wires, we taxied back to the cat. This was great. We would bag a lot of night traps at this rate. Night traps usually take a lot of time and fuel, but tonight we could get quite a few in a short amount of time. We OK'd the weight board, got into the shuttle, full power, lights on and were airborne, airspeed, gear up, climbing. We repeated this routine with the turn to base being reduced to 2 miles. Our fuel was 4.8 for our 3rd trap. We had been keeping track of our fuel and decided to let the ship know that we needed to be refueled after our fourth trap. Our Bingo[113] fuel requirement was 3.2 as our divert was Atsugi at approximately 150nmi to the North-East.

We trapped with fuel at 4.0 on our 4th trap and I called, "Eagle 507 is Trick or Treat[114]". Approach replied, "Copy, Eagle 507, Trick or Treat'. The call didn't seem to have an impact on the YS as we were taxied back to the cat for takeoff. We got airborne and repeated the short, nighttime pattern. KO flew another great pass and we trapped with 3.2k of fuel.

I called, "Boss, Eagle 507 is hold-down"[115]. By starting my radio call with "Boss", I was getting the attention of the Air Boss in the Control Tower. We wanted him to know that we needed to refuel before takeoff. The Boss replied, "Copy 507, hold down". We knew that voice, it was the Air Boss. He was supervising all flight operations and no one else onboard the carrier responded for the Air Boss. He knew our situation. We trusted that he was going to take care of us. However, the YS taxied us back to the cat. We were shaking our heads, 'No'. The YS saw our head motions, paused, spoke into his headset microphone, listened to the response and continued taxiing us to the catapult.

"What the hell is going on?", KO asked of me. We hoped the whole world would come to their senses for our predicament. We told the Boss of our situation. It would have been inappropriate for us to call the Boss again. It would be interpreted as helping him to do his job. That was not acceptable. We were taxied into the shuttle and awaited a command for full power. Certainly, the Boss would see what was happening and keep us on deck to get more fuel. That didn't happen. Full power was commanded, KO followed the direction and added full power, reviewed our instruments, Warning/Caution lights and turned his anticollision lights on. A few moments passed and down the deck we raced, getting airborne once again.

---

[113] Bingo climb profiles assigned climb speed/Mach number (M)/altitude to minimize fuel consumption for a specific distance.
[114] This call let everyone know that we could accept one more trap but then needed more fuel before the next launch.
[115] 'Hold down' tells the ship that the aircraft must be refueled prior to the next catapult.

Airspeed good, climb established, landing gear up and call, "Approach, Eagle 507 is airborne, we are Bingo"[116]. There was a pause on the radio. We had clearly caught someone off guard. It was hard to imagine as we had clearly communicated our fuel state many times and let everyone know we were Trick or Treat then Hold Down. "Copy Eagle 507, your divert is NAF Atsugi at 015° at 145nmi" came the reply. KO instantly added full power and brought the nose up to climb. I replied, "Eagle 507, Atsugi 015°, 145nmi, climbing to FL 410". KO retracted the flaps, configured the aircraft for cruise and set the pitch for 290KIAS until we hit M 0.70[117]. Once level at FL410, KO set power to maintain altitude and 207KIAS. That speed minimized our fuel usage. We crosschecked our range to Atsugi and fuel remaining. We predicted landing with 2000lbs provided other problems didn't occur.

We had time to discuss the events and couldn't figure out why we weren't held down for more fuel. We were disappointed that we couldn't bag more traps as it was so quick and easy to do with the help of the full moon. We were excited to be headed to Atsugi for the night as we had been aboard ship for over a week. Comfortable beds would be a nice change. In KO's case, he was happy to be returning to his spouse, Kirsty.

As we approached the Japanese Kanto Plain from the south, a third reason that made this night memorable came into view. The Plain is a relatively flat area that is slightly larger than the state of Massachusetts. The enormous population of the greater Tokyo area resides on the Plain. From 41,000 feet on that crystal clear night, the terrain below appeared like black velvet with millions of diamonds sparkling in contrast. The diamonds were the lights of the civilization below us. It was stunning. KO and I commented and stared. We had never seen anything like it. We enjoyed that view for about 20 minutes as we continued north. We contacted Yokota approach and were given an enroute descent to Atsugi, landing there at 1145pm. KO called Kirsty and she picked us up at the

---

[116] The 'Bingo' call essentially declared an emergency due to the fuel required to safely fly to NAF Atsugi. The fuel quantity was below carrier operations minimums. A reserve of 2K after landing in Atsugi required 3.2K to get there.

[117] A Mach (M) number is the ratio of an aircraft's speed to the speed of sound at a particular altitude. M1.0 at sea level is 661 KIAS, 761 miles per hour (mph) and 1224 km/hr. The speed lowers with altitude; https://www1.grc.nasa.gov/beginners-guide-to-aeronautics/speed-of-sound-interactive/

flightline. She was very surprised and excited about our unexpected arrival. The airwing followed us into Atsugi over the next few days and Midway pulled into port for Christmas. We had been given a gift of an early start on the holidays. Thanks, Boss.

Friday 19 December Our squadron had a Change of Command (COC) as CDR Cash relieved CDR Rhoades. Our new XO BN was CDR 'Polecat' Polatty[118]. He always seemed to be laughing enthusiastically about something. The JOs looked forward to serving with him.

Monday 22 December Beav and I were fortunate to fly an Intruder to Marine Corps Air Station (MCAS) Iwakuni to brief a Marine squadron on procedures and tactics of the Harpoon missile. We were our airwing experts on the weapon. Other squadrons needed help in digesting all the information and prepping their jets for use of the missile.

As the new year started, we flew out of Atsugi to keep our landing currency up to date then headed back down to NAS Cubi Point. A new senior pilot, LCDR Steve 'Smiley' Enewold, arrived at the squadron. He earned his callsign by always smiling. He was an Air Force Test Pilot School graduate that helped develop improvements to the Intruder software at NAS China Lake in the high desert of California[119].

Saturday 10 January 1987 Smiley and I flew a KA-6D down to Cubi logging a 5.5-hour flight with a fuel stop at Kadena AFB on Okinawa along the way. I shared FCLPs with Fig, Barney and KO in the following days.

Wednesday 14 January Barney and I flew aboard USS Midway from Cubi bagging a few traps with an hour of flight time. We refueled onboard and bagged another three traps. Ed and I had a night flight to regain his currency with a single night trap.

---

[118] Admiral Polatty later served as the Airwing One Deputy Commander during Desert Storm aboard USS America, Commander, Carrier Group One and the Commander of the Naval Training Center, Great Lakes, Illinois.
[119] Smiley was an inspiring mentor, leading to my application to Navy Test Pilot School and orders to China Lake.

Tuesday 27 January The CO and I had a one vs one Defensive Air Combat Maneuvering (DACM) flight with an airwing Hornet after launching from the Midway. The CO wanted to refresh his DACM skills while evaluating the skills of the other pilot and the maneuverability of the Hornet. We joined up with the Hornet shortly after launch and headed away from the carrier under the supervision of the E-2C Hawkeye aircraft. Once in a clear area, we flew away from one another for a few minutes and then turned toward each other for the engagements. Despite the CO visually acquiring the Hornet, turning toward him to pass nose to nose and pulling 6.5Gs to engage, the Hornet pilot was able to quickly secure a rear-aspect position. The Hornet's instantaneous turn rate and 7.5G loading capability, acquired an advantage on each pass.

As the Hornet pilot attempted to obtain a simulated gun's shot, he was taken by surprise. The CO flew the aircraft in an aggressive and erratic manner that prevented the Hornet pilot from securing a shot. Just as the Hornet pilot was settling in for a shot, the CO would change the dynamics requiring the Hornet pilot to reset and recommence a gun's approach. We conducted numerous engagements during the flight with nearly identical outcomes. It was quite exhausting yet exhilarating and gratifying to be making life more difficult for the aircraft with superior technology and flying performance. In the debrief, the Hornet pilot admitted his frustration yet admiration for our CO and his DACM skills. We had given him much more of a workout than he had anticipated.

Late January I had an idea to capture video of our airwing jets flying right after takeoff. I placed my VHS video camera in the starboard porthole in Midway's bow in the fo'c'sle area just before an aircraft launch sequence. Before long, I heard a jet at full power and knew the catapult launch was imminent. I heard the aircraft racing down the catapult toward the bow. The launch was from the port catapult and I missed the early moments of flight. I zoomed in on the Intruder as it flew away from the carrier. I captured more takeoffs of Intruder and Hornet aircraft. The video was rough but the idea had essentially worked. The original is posted on Facebook[120]. The ship's radar reflecting off the aircraft shortly after takeoff makes the video wiggle then clears as they get further away.

---

[120] USS Midway catapults;
https://www.facebook.com/david.a.maybury/videos/10218309194187637

Early February There has been an aircraft carrier near the Persian Gulf region almost continuously for decades. The carrier in the Arabian Sea in February was the USS Kitty Hawk (KH). Secret message traffic was sent from KH Intruder squadron, the VA-165 Boomers, to the Pentagon. They requested three more Intruders to augment their squadron aircraft to respond to Iran's attack of Kuwaiti tankers since May 1984. A message was sent to VA-115 in early February tasking them with providing three aircraft and crews to VA-165.

A plan was developed to have three Intruders launch from the USS Midway off the west coast of the Philippines, rendezvous with an Air Force KC-10 tanker aircraft and fly to Diego Garcia (DG)[121]. The tanker would provide fuel to the three aircraft every couple of hours during the approximately 9-hour flight. Our CO selected three crews and a spare crew if one of the jets had maintenance problems once airborne. The Eagle crews were the CO/Six Gun, Rocco/Jim, Smiley/Nubian[122] and HMAC/Gary in the spare. If one of the aircraft couldn't take fuel from the KC-10, the spare aircraft would take its place. As time passed, Jim got sick with a cold and I was selected to replace him.

Our squadron was quite excited to be involved in this tasking. The crews involved met in Aircraft Carrier Intelligence Center (CVIC) to prepare a brief for the admiral. I had been involved in using the new mission planning software, Tactical Aircraft Mission Planning System (TAMPS). It helped identify enemy defenses and vulnerabilities and facilitated flight planning. We used this tasking to brief TAMPS to our one-star admiral onboard the USS Midway. I spent days with the software preparing routes that would optimize the success of missions involving Iran. We used these routes in briefing the admiral on possible combat contingencies.

The evening before we left for Diego Garcia was rather busy. We needed to do a urinalysis for Gary as part of the squadron's periodic nuclear aircrew certification. The squadron Ordnance Officer (Gunner)

---

[121] DG is a coral island 1100 miles south of India and 2300 miles east of Tanzania, Africa.

[122] Nubian earned his callsign as the squadron hadn't had a new BN in a while when he showed up; he was the 'New BN'.

came into the ready room with six, 38 Caliber pistols and ammo. Smiley collected them all for safe keeping. I scurried about getting dog tags, my generic flight suit without the squadron logo from the ship's laundry and fingerprints for an authentication card. The CO did a practice brief in front of the participating aircrew, CAG Bowman and a few intelligence guys. It was short, sweet and to the point. The formal brief was given to the admiral adding a note that I was the only pollywog (person that hadn't crossed the equator onboard a Navy ship) of the aircrew going to the Kitty Hawk. The admiral seemed to enjoy that point. I didn't understand the importance of the comment.

Tuesday 17 February We briefed early in the morning. The plan looked simple enough. The weather was forecast to be great most of the way. After the brief, we waited for a 3-star admiral that flew aboard to wish us well for our journey. The USS Midway CO, Captain Mixson and CAG Bowman escorted the admiral into our ready room. He told us that he didn't know why we were headed to the KH but that someone obviously had a plan. We were surprised that he didn't know why we were going. Maybe he knew but couldn't tell us. We collected our overnight bags, the aircraft maintenance logbooks, put on our flight gear and walked to the jets on the flight deck. The Intruder had an area in the rear of the aircraft called the 'birdcage[123]' that allowed storage of our overnight bags.

Our four Intruder aircraft launched from Midway, quickly joined up and contacted the KC-10 on the radio. He was still on the deck at Clark AFB awaiting a takeoff clearance. Ten minutes later, we spotted him in the distance well below us as we had climbed up to a cruise altitude to conserve fuel. He appeared quite small at first then grew quickly as he approached. The KC-10 is a beautiful national asset! We joined up in trail as the KC-10 had the lead for the flight. He made all the radio calls for our formation. We conducted our first inflight refueling while near Midway in the event one of the aircraft had a refueling problem. If an aircraft had a mechanical or electrical failure that prevented them from receiving fuel, that aircraft was to return to Midway. Each of the three primary Intruders accepted fuel and filled their

---

[123] There was an access area aft of the engines that allowed storage of a bag of clothes for a cross country.

tanks. HMAC and Gary in the spare aircraft split off and returned to Midway.

The KC-10 was the lead of our formation of aircraft to DG. None of our crews had ever been there. We flew in a loose formation allowing us to engage the autopilot to relieve pilot fatigue. After reaching 16klbs of internal fuel, we lined up as a group on the port wing of the KC-10 to top up our external tanks. If there was a tanking problem, the internal fuel would allow us to reach divert fields with a sufficient reserve.

Our CO was instructed not to fly over any land during this mission. He communicated that to the KC-10 PIC before takeoff and he agreed to comply. This was an exciting journey that certainly topped previous ones! Singapore was very pretty to see and was larger than I expected. As we made a left turn toward DG, I made my last INS position update using the radar cursors on Banda Aceh[124]. No velocity correct was needed as my INS was working well. A well-performing system makes life easier in combat, so I was pleased. Land passed off our left side and vanished behind us. Water was our companion for the many hours ahead.

We finished our last tanking evolution with the KC-10 and brought the speed up, traveling at over 500KGS. DG appeared on the radar and the INS had drifted very little throughout the flight. As we got closer to Diego, we were impressed as it was quite beautiful! We came into the break as a three-plane and pulled 5Gs into the downwind. The KC-10 trailed miles behind. As Rocco and I were third in the formation, we could see the first two aircraft kicked up coral dust from the runway as they landed. We taxied and shutdown and were immediately met by British customs agents. They announced that no firearms were allowed on the island. We asked if that included our personal survival 38 caliber pistols. A lot of discussion took place and the CO promised to take personal responsibility for securing the weapons throughout our overnight stay. The agents agreed and let him keep them. The KC-10 landed and our squadron maintenance crews promptly appeared from the tanker to start working on our jets. They really appreciated the ride in the KC-10, watching us refuel from a rear portal window and being part of

---

[124] Banda Aceh was the location of the horrific tsunami on 26 December 2004 killing an estimated 230,000 persons: https://www.history.com/news/deadliest-tsunami-2004-indian-ocean.

the adventure! Their enthusiasm and broad smiles echoed their soaring spirits.

We got a ride to our accommodation with explanations from an S-3 LCDR NFO. Three S-3s from KH were flown to DG to make room for our three Intruders. Unfortunately, there was no room at the BOQ. The barracks had vacancies so we slept there, three crew to a room. This was the Navy way of life. The rooms were clean and adequate. We all needed to stretch our legs after all that flying, so we happily walked over to the Officer's Club. Stories of our journey were shared over a beer. An oriental buffet provided a delicious dinner. The KC-10 pilots came in for a soda and we shared greetings. They were great guys and we were very grateful for their help in getting us to DG.

Wednesday 18 February Everyone arrived at the brief, looking bright-eyed and happy to continue the journey. We discussed the flight north to the KH in the Arabian Sea. It was a five-hour flight so we needed a couple of inflight refueling cycles to safely get there. We toured inside the KC-10 aircraft, had discussion with our maintenance guys and started engines. We were so proud to be working with the KC-10 crews and sensed a mutual admiration from them. The KC-10 took off first and orbited overhead Diego waiting for us to join before we pushed north. We took off as a flight of three with 10-seconds between each aircraft releasing their brakes. The 10-second stagger between aircraft enhanced safely in the event an aircraft experienced a problem during the takeoff roll.

The three Intruders joined up quickly then joined the KC-10 that was headed north in the climb. The flight toward KH was uneventful. We found that a shallow descent gave us more gas during refueling. We turned east for our final refueling with our aircraft breaking north when complete. The KC-10 continued non-stop to Guam![125]

We tuned our radios to the KH frequencies and checked in with Strike. They passed us over to Approach and Tower. We hit the break at 800 feet with 450KTAS and turned downwind with a 5 second separation between aircraft. We hit our numbers at the 180, 90, 45 and rolled into

---

[125] Guam is 1600 miles east of Manila, Philippines. The KC-10 trip after tanking us would have been 5000 miles!

the groove. 'That is a big deck' was my first thought. Rocco flew a great pass and we came to a thundering stop in the arresting gear.

Once out of the gear, we started taxiing toward the bow. It was like a shopping mall parking lot while Midway was like a few parking spots in front of a convenience store. They could fit a whole six-pack of aircraft ahead of the super-structure and behind the bow Jet Blast Deflector (JBD)s. A vehicle painted like an Air Force follow-me truck brought us to our parking places. A follow-me truck on an aircraft carrier was unheard of! We suspected it was a joke at our expense but the paint job was quite nice. We shut down the jets and were met by smiling Boomer crews to welcome us aboard.

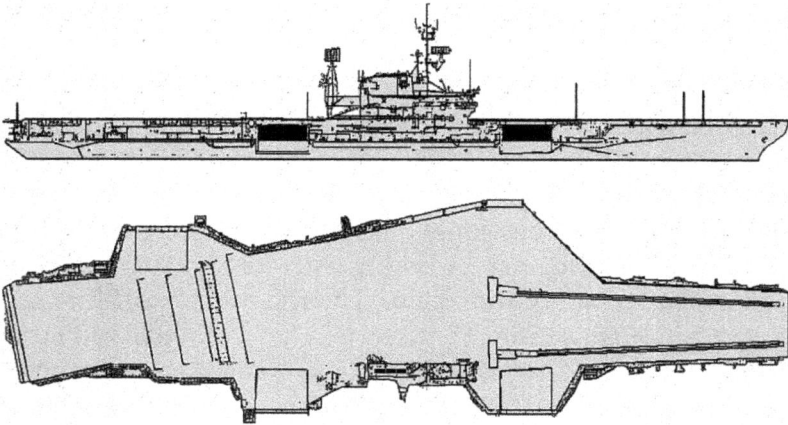

The USS Midway depicted above had three arresting gear, two catapults, three elevators, a length of 979 feet and flight deck width of 238 feet. Despite its smaller size, Midway Magic delivered excellence.

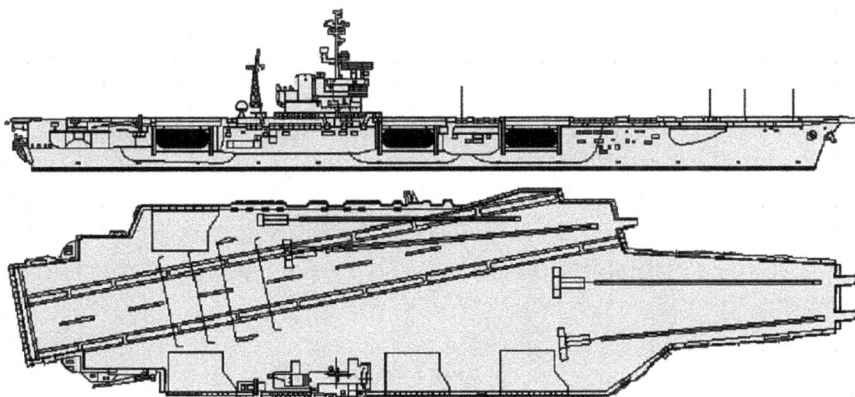

The USS Kitty Hawk with four arresting gear, four catapults, four elevators, length of 1063 feet and flight deck width of 252 feet. Note the greater distance from the superstructure to the forward catapults.

We traveled down to maintenance control, handed over the logbooks for our aircraft, debriefed a few minor gripes and headed over to the Boomer Ready Room. The VA-165 CO, callsign Scrapper, greeted us upon our arrival. He was beaming as if it was Thanksgiving, Christmas and his birthday all rolled into one. His squadron had just borrowed three Intruders from another squadron that was located off the coast of the Philippines. It was a highly unlikely achievement. Scrapper promised to brief the reason for our long journey later that evening along with a flight demonstration. He suggested we find our racks, food and relax.

As promised, Scrapper showed us the contingency plans that had been developed. Iran had been harassing Iraq and Kuwait for a while. There was concern that their misbehavior might escalate requiring response from the USA. As the plans were explained, we could see the hand of a master tactician at work. The only problem with the elaborate plan was that more aircraft were needed to reduce the risk. Risk reduction prompted their request to the Pentagon and our tasking to move the three aircraft to the KH. As we were Eagles working with the Boomers, we gained the nickname of Beagles. It fit and we were honored to be there.

Following the brief, Scrapper took us up to the Tower to watch a night launch sequence. Aircraft were launched from the bow cats only. Intruders launched from Cats 1 and 2 with ten seconds separation

111

between launches. This allowed the second aircraft to quickly join on the aircraft in front. The next two aircraft were launched two minutes later with a ten-second separation. The first two aircraft were at 250KIAS, joined up, in a port circle and 180 degrees out for the launch of aircraft three and four. Three and four joined up, looked up and in front of them was the formation lead enabling a simple rendezvous. This continued until all the strike aircraft had joined the formation. It was a very time and fuel-efficient manner of collecting a strike package from the ship. We were impressed!

Friday 20 February Rocco and I started flying from the KH. As time passed, we secured Rocco's night currency and practiced contingency plans.

Tuesday 24 Feb As we waited for an Iranian transgression, there was a request for KH to travel south and maintain a presence off the coast of northern Somalia. Coordination with Somali government agencies approved low level overflight of the nation under strict guidelines. Rocco and I entered Somali airspace at a preassigned time and place and similarly departed the airspace at a specific time and place. That was much more restrictive than we were accustomed, but it enabled us to descend to 200 feet AGL for a thirty-minute low level.

The terrain was desolate with little foliage or water. We flew near several herds of animals that appeared to be sheep. As we neared the departure point of the low level, we were required to climb to 1000 feet AGL and reduce our power to achieve 300 KTAS. This was required to reduce noise around a small village. As we approached the village, a remarkable view was ingrained in my memory. There was a well in the center of the village surrounded by a couple dozen buildings. Dirt paths extended radially away from the well in all directions and beyond the horizon. The horizon was 38 nautical miles (70 kilometers) from our vantage point. I could not see anything resembling a hut or a village between the well and the horizon along any of the tracks. That well served a widely spread-out group of people. There were about 50 people around the well. How could they survive with such a long journey for water? We departed Somali airspace and returned to the KH.

As we spent time with the Boomers, various personalities emerged in the squadron. LCDR Jeffrey Wieringa served as the Tactics Officer for the squadron. This was a department head billet for the Boomers. This was unusual as most other squadrons gave the job to a Lieutenant. The Boomers were serious about tactics and his expertise showed in their combat plans. Jeffrey was a Navy Test Pilot School graduate with a brilliant mind. He knew details about our aircraft that no one else knew. Occasionally he would share a portion of his knowledge during AOMs. The JOs would listen for a few moments then nearly on cue would start in unison, "We can't hear you Jeffrey, Na Na Na Na. We can't hear you Jeffrey". That was their way of telling Jeffrey that they didn't understand what he was saying, didn't want to be confused by what he was saying and would appreciate it if he would stop talking. That was funny to watch. Jeffrey was trying to give everyone the benefit of his advanced insight into aircraft and tactics[126]. Some of the JOs weren't ready to learn what he had to offer.

27 March A crew swap occurred with the CO, Rocco, and Six-Gun leaving, pilots Aldo and K-9 and BN, Bob taking their places. Aldo became my pilot for the reminder of the flights aboard KH.

Aldo and I were standing an Alert 15 tanker duty in full flight gear in the Boomer Ready Room. The Alert 5 Tomcat was called for launch over the ship's 1MC due to an incoming Soviet bomber. I immediately stood up to walk to our tanker for takeoff. That is what we did aboard Midway when the Alert Five fighter was launched. The Hornets, or the Phantoms before them, always needed fuel from a tanker for a bomber intercept. One of the Boomer aircrews asked me, "Where do you think you are going?" I replied, "I'm going to launch the Alert Fifteen tanker". He replied, "They called away the Alert Five fighter. They didn't ask for the Alert Fifteen tanker. The Tomcat can do that mission without any extra gas. Take your seat". I was astounded! I sat down but expected the Alert Fifteen tanker to be called for takeoff at any time. I waited for a couple of hours then the Tomcat finally landed after intercepting the bomber. The Tomcat had sufficient fuel to do that mission autonomously. That was impressive and much less of a burden to the carrier and airwing!

---

[126] In 1996, Jeffrey and I flew together in Hornet chase aircraft during the flight test of the Super Hornet.

Tuesday 14 April Aldo and I had day and night flights to regain his landing currency after a five-day stop in Karachi, Pakistan. We had six more flights in the following ten days.

In late April, as the Iranians were staying quiet with their neighbors, the KH was allowed to steam south to make a port call in Mombasa, Kenya on the east coast of Africa. As the KH's time in the Arabian Sea was ending, it was decided that the three Beagle crews and their aircraft could depart and head back to Midway. There was one more important item of business that needed to be sorted out before we could leave.

Saturday 25 April The day had arrived when pollywogs were initiated into the Realm of Neptunis Rex and became 'Trusty Shellbacks' onboard KH[127]. I finally learned why the admiral onboard the USS Midway was so pleased to hear of my need to be 'cleansed' during this deployment. As a 'slimy pollywog', crossing the equator required 'indoctrination' to become a 'Trusty Shellback'. The Navy had a long-standing process that required pollywogs to demonstrate strength under adversity to prove themselves worthy to be called Shellbacks. A group of Boomer / Beagle aircrew pollywogs knew the indoctrination would be a challenge but decided to add spice to the process. Instead of sleeping in our normal staterooms, we all met in one stateroom at 4am on the morning of our initiation. We locked and barricaded the door with heavy steel footlockers stacked from the deck to the overhead. We waited quietly for the festivities to begin.

At 7am, there was a knock on the door and the lock was checked. We remained quiet and those at the door went away. At 745am, there was another group knocking on the door and trying to enter the room. We stayed quiet and they also departed. At 815am, a group arrived pounding on the door saying, "All right, you slimy pollywogs, we know that you are in there. Open the door or there will be trouble". We stayed quiet but were all smiling as we were really enjoying this. "If we must tear down this door, you will be paying a big price. Open the door!" was yelled as they continued pounding. Boomer pilot and pollywog Dish replied in a

---

[127] https://www.history.navy.mil/browse-by-topic/heritage/customs-and-traditions0/crossing-line.html

high pitched, woman's voice, "Who is it?" "Open the door, you slimy wog" came from outside the door. Without missing a beat, Dish replied in his high-pitched voice, "You will have to wait. We're getting dressed". That really made them angry. "You asked for it, wogs" was the response.

The Shellbacks started pounding on the door hinges with a sledgehammer. That was loud! We didn't quite expect that ingenuity or intensity. They beat on the door for about 10 minutes and smashed it off its hinges[128]. Once the door was off the hinges, the Shellbacks removed the footlocker barricade by pulling them into the passageway. Once a path was clear, a big storm of Shellbacks came into the room and dragged the pollywogs out and down to the Boomer Ready Room. Life as we knew it suddenly changed.

The Boomer Ready Room had been transformed into an 'Indoctrination Center' with the chairs pulled to one end and a tarp placed in the center of the room. Flour was placed on the tarp. We were covered in oil then rolled in flour to coat our bodies. We received further training in the ready room then were escorted on our hands and knees down passageways and up to the hangar bay. It was an assembly line of pollywogs receiving further training and refining in naval tradition. After spending over an hour on the hangar bay, we were escorted up to the flight deck for more advanced training. We were provided the rare opportunity to crawl through narrow tubes of garbage to obtain higher levels of consciousness. After another hour of training on the flight deck, we gained much clearer insight into worldly things and became Shellbacks.

There were showers constructed on the flight deck to enable us to rinse off the crud that accumulated on our bodies throughout the day. We removed and disposed of our clothes and went down below decks for hot showers. It felt good to be clean and wearing fresh clothes. The Shellbacks had done a great job of executing a large-scale indoctrination of over 2000 pollywogs. It was an impressive operation!

---

[128] The Executive Officer of the Kitty Hawk was furious when he heard that the door had been pounded off its hinges. He demanded its immediate repair, and it was promptly restored to nearly new condition.

Sunday 26 April The Beagles departed the KH and flew back to DG to await our KC-10 tanker for transit back to the Philippines. Much to our great surprise, there was not a KC-10 waiting for us, nor was there one to support us for another week! As we had no control over our situation, we decided to make the best of it and DG had great ways to do that. We played volleyball every day to stay fit. The games were long and hot due to DG's proximity to the equator. They had outdoor movies every night in a courtyard area. The O' Club served great meals and the beer was cold.

Sunday 3 May We finally departed DG on the wing of a KC-10. Smiley/Nubian were in our lead aircraft with K9/Bob and Aldo and I as wingmen. We refueled every couple of hours as we had on our previous flights. As we approached the Philippines, we topped off our fuel and said goodbye to our airborne gas station. The KC-10 crew added power and accelerated briskly away from us. We were impressed at the acceleration considering the size of the aircraft! We had been cruising at M0.75 but the KC-10 cruised at M0.85 for efficiency. We landed at NAS Cubi, caught up with our squadron mates and had a great reunion! Our excitement grew in the following weeks as the USS Midway was headed to Australia!

We flew a couple days out of NAS Cubi then boarded the ship to sail south. I flew primarily with Smiley at this point. He had a perpetually joyful personality. As Smiley had worked on the Intruder mission computer software development at China Lake, he knew a lot more than our other pilots (and many BNs) about how the software in the jet worked. One day I said to him, "Flying with you is like having a pilot and two BNs in the cockpit".

Monday 25 May As we sailed south of the Philippines, there was a big cleaning job to do aboard ship. There were many slimy pollywogs aboard that needed to be cleansed into Trusty Shellbacks. I'd experienced the cleansing aboard the KH and was now equipped to help my fellow squadron mates. We had a meeting of all Trusty Shellback aircrew in our squadron. It was decided that each pollywog should be given a phrase to memorize and recite verbatim whenever a Trusty Shellback might request it. It was important for the pollywogs recite these phrases regularly as part of their cleansing. Some thought was given to each individual

116

pollywog to determine the best phrase for each one. When it came to KO, the Shellbacks agreed that his phrase would be, "I know, that you know, that I know everything". Other pollywogs were given suitable phrases to match their personalities as well.

As we headed south towards Sydney, coordination occurred between our airwing and the Australian Air Force. An exercise was scheduled in which the Australians would simulate an attack on the USS Midway. The Australians were expected to fly out of Royal Australian Air Force (RAAF) Base Townsville on the Queensland coast. Our airwing set up a 4-plane of F/A-18 Hornets in a CAP between Midway and the coast. As Midway was 400nmi east from the coast, our CAP position was 200nmi to the west toward the coastline. We supported our Hornets with tanker fuel and they rotated crews while we awaited the attack. As there was no time scheduled for the attack, we had to be ready at any moment. The E-2C aircraft was airborne and ready to vector the Hornets to the inbound attackers.

A CAP had been in place for eight hours without any activity at all. Suddenly things got very exciting as the E-2C picked up a low altitude contact moving at supersonic speed from the EAST toward Midway! The CAP was ordered to reposition to the east to counter the threat. Unfortunately, there was no possible way for the Hornets to reach the inbound threat in time. The USS Midway had lost the exercise as the Australians outsmarted us. They had launched an F-111 aircraft that flew a very long track to attack the Midway from the east to completely avoid our CAP positions. We didn't notice the attack until it was far too late. While it was embarrassing for our airwing, it was a testament to the ingenuity of the Australians and the range and speed of the F-111. It was truly an amazing warfighting machine![129]

Friday 5-15 June The USS Midway pulled into Sydney Harbour on 5 June to celebrate the 45th Anniversary of the Battle of Midway in

---

[129] The RAAF retired their F-111 fleet in December 2010. Thirteen of the aircraft were given to museums while twenty-three were buried at a landfill outside of Amberly airbase due to the asbestos construction of the fuselage.
https://theaviationgeekclub.com/heres-why-australia-buried-23-f-111s-after-the-aircrafts-retirement

WWII[130]. Midway's crew was very excited to be there. Those crewmembers that had been to Australia before had only great things to say about it. There was a large group of boats in the harbor to welcome us. We tied up to a buoy in the harbor and boats delivered us to the Woolloomooloo Pier, not far from the Opera House.

I was given the task of securing a large suite at one of the better hotels in Sydney to serve as a party venue for our Eagle officers. Officers paid $100 for a suite supplied with food and drinks. Bing and I found the very nice Hyatt hotel at the top of Kings Cross[131].

After ten days in port, we settled our account with the Hyatt and made our way back to the Midway. It had been a great time in port with everyone enjoying themselves and loving Australia. We didn't really want to leave but duty called.

---

[130] Our visit was the first carrier to visit Sydney since 1971.
https://www.navysite.de/cruisebooks/cv41-88/005.htm
[131] Kings Cross is a popular party location with many pubs, discos and nightclubs.

# Chapter 7

## Midway Rolling, Harpoon, Kuwaiti Tankers

*The Iranians had previously used Silkworm missiles launched from their shores to attack the tankers. We knew the location of the missile launch sites. Our Intruders bombers were loaded with four MK20 Rockeye II Cluster bombs to destroy the Silkworm missile sites if they launched toward the convoy.*

As we sailed further north, the Midway started rolling back and forth more dramatically than ever experienced. The rolls were over 15° which was large for such a big ship[132]. As we were conducting flight operations, the roll became a new challenging part of flying from the Midway. We felt reasonably comfortable while our aircraft was chained to the deck of the carrier. The rolling was so significant that our view of the sea was blocked while sitting high above the flight deck in our jets. We stared down the flight deck at a menacing sea as the roll reversed. We armed our ejection seats just in case we left the flight deck. After engine start, the chains were removed just before our taxi to the catapults. Taxiing required careful attention to detail as if we were flying. The emergency brake was set but with the big rolls, we weren't exactly certain if it would contain our movement.

Once ready to taxi away from the carrier deck edge, the YS signaled to release the emergency brake off and add power. The YS timed this event so that the aircraft would roll downhill into the center of the carrier and away from the deck edge. The pilots added power to taxi into the center but added brakes again as our nose started to rise above the horizon. We didn't want to add power to fight that roll angle as it was futile. Additionally, we could be spraying sailors behind the aircraft with high powered jet exhaust. The pilots held the brakes until the nose came down then started the taxi again. They were mindful to ride the brakes to prevent the aircraft from racing downhill to the other side of the carrier. A sigh of relief took place after we turned the aircraft towards the catapults and didn't need to fight the sea-saw of the rolls. This process became the norm in time, but it was always performed with care. I flew

---

[132] On 8 October 1988, the USS Midway rolled 26° in the middle of the night sending the ship into General Quarters.

from Midway with several different pilots during these rolls and they all managed them with the utmost professionalism. Without the expertise of the entire airwing and flight deck personnel, flight operations could have easily been deemed too dangerous and cancelled during these rolls. The Midway professionalism or 'USS Midway Magic' as we called it, brought us through difficult moments.

Wednesday 1 July Beav[133] and I launched a Harpoon[134] missile from our aircraft scoring a direct hit on a target hulk. The success was the culmination of a considerable amount of work by many different departments. VA-115 found out that they were going to fire a Harpoon months before the event. Before ever reaching the launch date, we needed aircraft that could fire the weapon. During the initial assessment, the AQs worked feverishly checking out the wiring in the jets with the assistance of onboard civilian Grumman Aerospace Technical Representative, Mr. Johnson. Only a couple of our jets could fire the Harpoon due to wiring problems. As it was a weapon that we had never fired before, no routine maintenance had been completed on the Harpoon specific wiring on our aircraft. Corrosion damages aircraft wiring and electrical connectors when flying from an aircraft carrier. The shop leader, AQ1 Mike developed a diagnostic tool to troubleshoot the wiring. They used the tool on all the bomber aircraft to determine if they showed promise for firing a Harpoon. Several of the aircraft were better than others and the AQs improved them in preparation to fire the missile.

We planned to launch the missile at near its maximum range to validate the capability. We were given an exact time for the Harpoon to strike a target hulk in open ocean off the western coast of the Philippines. We launched from the Midway in the Philippine Sea and positioned ourselves to fire the Harpoon. An F/A-18 aircraft flew on our wing to serve as a chase aircraft to follow the Harpoon from the release point to the target verifying impact. The missile had a telemetry package onboard allowing it to be flown into the ocean if there was a malfunction. The Hornet pilot was the safety observer that ensured no inadvertent flight

---

[133] Rear Admiral 'Beav' Horton had a successful career in the Navy; https://www.gettyimages.com.au/detail/news-photo/rear-admiral-ron-horton-commander-of-task-force-73-and-news-photo/103322697
[134] The Harpoon was our primary anti-ship weapon; https://www.NAVAIR.navy.mil/product/Harpoon

deviation occurred. We arrived at the release point, the missile came off and our wingman gave chase to the missile until it impacted the target.

After landing aboard Midway, we taxied to a parking spot and shut down the jet. The maintenance crews congratulated us on successfully firing the missile. When we first started, the squadron didn't even have an aircraft that could shoot a Harpoon. We had just taken a long-range Harpoon shot and achieved a bullseye. That was something to celebrate!

June A new pilot, Waterdog (Dog, for short), showed up while we were headed back north to Atsugi. He and I were crewed together for 45 flights from June to November 1987. Dog was a young, energetic young man with a boyish charm and a big smile. He looked too young to be a fleet Intruder pilot. Flying with Dog was relaxed due to his excellent skills. Our first flight was from Midway during the day. I recall Dog's OK three wire landing. That was a great start. Successive flights from the Midway were similar. He landed with many OK, three wires. Watching him land was like poetry in motion and a pleasure to experience.

Mid-July 1987 The USS Midway arrived back to Yokosuka. We had been away from home for six months. Those of us that didn't make the cut to fly off the ship, made our way via car to Atsugi.

August 87 Dog and I had a few day and night flights from Atsugi. We were building up our crew coordination while getting ready for a CAS detachment to Daegu in South Korea. CAS uses air firepower to assist troops on the ground. The attacking aircraft is controlled by a Forward Air Controller (FAC) on the ground or airborne. Our aircraft would hold at prebriefed safe locations that were controlled by our troops then sent into areas held by the enemy.

Our CAS detachment was conducted in close coordination with a USMC unit. The terrain was mountainous, we were flying at low altitude and at high speed for our attack runs. The Marines did a great job in communication and designating the targets enabling our attacks. Dog and I enjoyed working with them. Flying from Daegu was interesting but not as intense as Suwon; we were a long way from the DMZ. There were still manned anti-aircraft guns around the perimeter of the field.

After our South Korean Detachment in Daegu, we flew from Atsugi for a month, were back aboard USS Midway for a couple of weeks of local operations near Japan then back to Atsugi for a last couple of weeks before a six-month cruise to the Middle East.

Friday 18 September Pilot Denny 'Seadog' Seipel and his BN Healy died in an A-6E Intruder while conducting a night, Case III approach to an aircraft carrier. Denny had been an Intruder BN and transitioned to Intruder pilot after a couple of years back at the training command to earn his pilot wings. He was in Bing's class at VA-128, had an engaging personality and was a leader in the Intruder community. Their deaths were a tragic loss.

Monday 21 September I completed my Mission Commander qualification with Smiley on a WASEX with a Balzam AGI[135], Soviet Intelligence collection ship. The aircraft involved in the WASEX were two Intruder Harpoon shooters, a Prowler, a KA-6D tanker, two Hornet HARM shooters and an E-2C for coordination. I briefed the mission, called the shots airborne and debriefed the flight after landing. The EA-6B provided critical help to complete the mission.

Wednesday 23 September VA-52 CO pilot, Lloyd Sledge and his BN John launched from the USS Vinson at night to return to NAS Whidbey Island. Lloyd suffered a hypoxic episode and passed out while at high altitude above the Pacific Ocean. John, a VA-128 classmate of mine, was severely injured during his ejection from the aircraft but survived. Lloyd died in the accident as the Intruder ejection seats worked independently from one another. This accident was another significant tragedy for the entire Intruder community.

Thursday 15 October K9 and I flew an Intruder aboard Midway as she started sailing south toward the Philippines for a six-month deployment to the Arabian Sea.

Wednesday 21 Oct Waterdog and I were fortunate to be scheduled for a night, low altitude practice attack on the NAS Cubi Point Officer's

---

[135] Balzam AGI ships often trailed the Midway battle group; https://en.wikipedia.org/wiki/Balzam-class_intelligence_ship

Club. We were part of a stream raid designed to simulate delivery of multiple loads of bombs on a high value asset. The Cubi runway was oriented 070° magnetic. The Officer's Club was positioned in the hills overlooking the runway. A 070° inbound attack heading from the southwest allowed us to stay low over water then clear the modest terrain around the target. There was higher terrain to the east and north but we planned to avoid it. Our attack profile at 500 AGL and 480 KGS simulated dropping snakeye bombs. Once the bomb release occurred, we planned a 6G left, 180° level turn to escape the frag pattern and line up with our departure heading out of Subic Bay.

Our target time was 10pm so we took off from Midway at 9pm. We joined with the other aircraft participating in the attack. Our separation was assured by our flight paths and one-minute intervals to ensure we didn't frag our own aircraft. We checked in with NAS Cubi Point approach and tower to ensure we had clearance to conduct this highly unusual flight profile.

I updated our navigation system as we approached the coast. Events happened very quickly in attacks with sometimes just seconds to hand off from the radar to the targeting FLIR for the final aimpoint. Our simulated bomb release point was less than 1nmi from the target. We were traveling at 8nmi per minute (480KGS). Dog was monitoring our progress by looking outside the windscreen while my head was buried in my radar/FLIR boot. I found the Club on my radar and slewed my cursors slightly right to update the targeting. As I did this, Dog's gunsight aimpoint slid slightly right as did the steering on his ADI. I was anxious to do final targeting with the FLIR, but the target wasn't showing. It was obscured by hills and trees.

Finally, the Club appeared on my FLIR. I transferred computer targeting by depressing the FLIR button on my slew stick and releasing it on my desired aimpoint. Once again Dog's targeting symbology shifted, he flew the new commanded heading. The release symbology marched down on the ADI signaling that weapon release was imminent. Once the simulated release was complete, Dog executed a left, 6G turn as I selected navigation to the egress point. He kept the throttles at military through the turn and egress. It was an exhilarating experience to be racing through the Cubi airspace at low altitude at night. We departed Subic

Bay, slowed down, climbed up, contacted USS Midway Strike and recovered aboard.

Tuesday 27 October Waterdog and I were scheduled for a day takeoff and night arrested landing aboard Midway. It started out as a normal flight but turned out to be rather challenging. When Dog arrived at Midway, his landings were great. His approaches to landing were smooth and seemed easy for him. I often sat next to him and didn't say a word as he responded to all flight profile variations appropriately. Dog gained a reputation as one of the best pilots for landings in the airwing. As an LSO, this was a badge of honor as he was debriefing other pilots on their landings. This night was different.

As weeks passed from his arrival to Midway, Dog's approaches to landing became more 'colorful' inside of a nautical mile to touchdown. For example, everything was going well until thirty seconds prior to landing then he would go too high then right then low then left. It was strange. He arrived to Midway with an uncanny ability to fly beautiful approaches to landing. Dog became unable to hold a steady flight path in close. I wasn't sure how to help him return to his previous steady hands. The other LSOs provided advice but nothing seemed to help.

After flying our mission, we set up in the marshall stack waiting for our time to commence the approach for our night landing aboard Midway. We planned our fuel to arrive at 6.0 for our trap. The approach was smooth yet Dog started a three-dimensional dance inside of a mile to touchdown. Most nights, he was able to settle down the variations and collect a wire with our arresting hook. It didn't happen on this approach and Dog applied full power as we flew away from the deck back into the dark night. He raised our landing gear with a positive rate of climb away from the water. We had boltered - our first one together. Bolters are an embarrassing moment for the carrier pilot as it places additional hardship on the already tired flight deck crew and all others working air operations. We knew that but there was nothing we could do but fly the aircraft and come back for another landing.

Approach called, "Eagle 501 turn left to a downwind heading of 195, climb and maintain 2000 feet". I repeated, "Eagle 501, left to 185, climbing to 2000 feet". We needed to shake off whatever just happened and focus on the task at hand. We arrived on heading 185, altitude of 2000 feet. We didn't need to talk about what happened. These things happen and we had seen enough of Dog's recent landings to know a bolter was possible. We waited for approach to give us our next direction. "Eagle 501, turn left to heading 095 for your base leg", came from Approach. "Eagle 501, left to heading 095", I replied. We were going to be on a 12 mile straight in approach. That was plenty of distance to feel comfortable and not rushed on our approach. Approach called, "Eagle 501, turn left to heading 352, dirty up and descend to 1200 feet when comfortable on the inbound leg". I replied, "Eagle 501, turn left to 352, dirty up and descend to 1200 feet on inbound leg".

Everything was feeling good. We had both done this so many times before. We received Midway's ACLS needles to help guide us on our approach. Approach called, "Eagle 501, report needles". I replied, "Eagle 501, fly up and left". "Eagle 501, good needles" replied approach. This indicated that we could rely on them for our approach navigation. Dog was good at flying the needles, making small corrections to keep them centered. We arrived at 3 nautical miles and started our descent to the carrier at 700 FPM. We saw the carrier with its lights against the dark sea but Dog focused on the needles to fly the approach. Using visual cues for an approach to a carrier at night has resulted in many tragedies over the years. Visual misinterpretation is commonplace when looking at lights on a dark sea[136]. The needles gave reliable information and were much easier to fly. "Eagle 501, ¾ of a mile, call the ball", came from the LSO. I replied, "Eagle 501, Intruder ball, 5.0". The LSO responded, "Roger ball, winds at 23 knots".

As we called the ball, Dog took over flying the aircraft visually using the meatball for glideslope information and the runway centerline lights for horizontal lineup. His start was smooth but then drifted back into a 3-dimensional dance. We hit the deck, Dog added full power and we were flying once again after a few seconds. Our second bolter of the night was not a good sign. Dog established a positive climb away from the water while raising the landing gear. We both took a deep breath and sighed. Approach called, "Eagle 501, turn left to a downwind heading of

---

[136] One dark night, while on a 5nmi final approach to the carrier, I sensed that our aircraft was in a 30° dive despite our instruments verifying that we were in level flight. The illusion remained for 60 seconds until we got closer to the carrier.

185, climb and maintain 2000 feet". "Eagle 501, turning left to 185, climbing to 2000 feet", I repeated.

Flying the aircraft was our first duty, so we did that. In the back of our minds, we wondered about the cause of the two bolters. Everything looked solid until we called the ball. Waterdog needed to smooth out his control inputs as he came close to the ship. That was easy to say but not so easy to do when things weren't going well. I felt incapable of helping Dog with the errors as his movements were too erratic to keep up with verbally. The bolters were terribly frustrating for Dog, especially after a great history of excellent landings aboard Midway. This was like a slump. It happens in professional sports and the critics love to discuss the athlete and his or her, 'problem'. There was no time for that kind of discussion which might shake the confidence of this young pilot. We needed to get aboard and put the bolters behind us. We discussed the approach briefly on the downwind leg. I tried to be encouraging citing that the approach was excellent until just in close then it got a bit erratic. Dog already knew that. He agreed, took a breath and we resumed the task of getting the aircraft back on that carrier deck.

The next approach was a repeat of the previous one. The carrier was trying to make this as easy as possible. "Eagle 501, ¾ of a mile, call the ball" came from the LSO. "Eagle 501, Intruder ball, 4.0", was my response. "Roger ball Intruder, winds at 25 knots", came from the LSO. We were on glidepath with good lineup when we called the ball. That changed as the approach continued. Dog applied full power as we touched down and yes… once again we were airborne. A positive rate of climb and our gear came up. This was not good at all. Approach called, "Eagle 501, your divert is NAS Cubi Point at 165 degrees at 150 miles. Climb and maintain 35,000 feet. Your tanker is at your one O' Clock and three miles". I replied, "Eagle 501, copy divert Cubi Point, 165 degrees, 150 miles, climb and maintain 35,000, tanker in sight". Everything had just changed. We weren't going to land on the carrier tonight. We were going to Cubi. We all loved Cubi but didn't want to go there under these circumstances. This was embarrassing.

The tanker was another Intruder, heading 165 degrees and climbing as well to 35,000. Folks down on the carrier had been planning this for a while and talking to the tanker crew. They didn't mention it to us as they didn't want to distract us from our task of landing the aircraft. Once we boltered for the third time, it was time to execute their plan. We had less than 4000lbs of fuel and were using it quickly as we climbed at our most fuel-efficient airspeed of 290 KIAS. We looked at the distance of 150 miles to Cubi Point. We anticipated landing above the 2000lb squadron SOP minimum fuel for land-based operations.

Dog joined up on the tanker's left side and waited to be cleared in to tank. The tanker's hose was already deployed for us. We were cleared to tank and slid in behind the tanker's hose. Normally, we refueled in level flight but we needed to keep climbing to conserve fuel if we couldn't receive fuel from the tanker. Our squadron pilots were quite comfortable tanking from other Intruder aircraft. Generally, the refueling basket was quite stable and easy for the pilots to engage with the probe on the first attempt. Dog added power once stable behind the basket. As we moved in close to the basket, we noticed that it was moving about in turbulent airflow. Dog needed to add finely timed aircraft control movements to compensate for the basket movement. Unfortunately, he miscalculated the basket's movement and we missed the first attempt at engagement. Dog reduced power to return behind the basket and set up for another approach.

Once stable, he added power and moved toward the basket. Again, the basket made unexpected movement and our probe slid past it. Dog reduced power again to reposition himself for another attempt. This wasn't good. Dog struggled coming aboard the carrier and now was struggling getting into the basket. He was frustrated with himself. I provided encouragement hoping to calm Dog with my words. He just needed to relax and get rid of the tension that was driving his control movements. As we spoke, he did calm down, missed the basket once more then got in. We were down to 3200lbs then started receiving fuel from the tanker. The green light from the tanker package confirmed that it was passing fuel to us. Hallelujah! We were still climbing through 25,000 feet. After a few minutes, approach called, "Eagle 501, now that you are receiving fuel, how would you feel about coming back to the carrier to get aboard?" I looked over at Dog. He nodded in the affirmative. I replied, "Approach, 501, that would be good. We have another 5 minutes of tanking to complete then we will be ready for a vector".

Dog was flying well. He was holding position as we increased our main fuel tank to 7000lbs to enable our descent, approach to landing and show up on the ball with 6.0. It was good for Dog to just fly formation and gain back confidence while getting our fuel. Once we were at 7.0, he reduced power, slid our probe out of the basket then sidestepped to the right wing of the tanker. Dog turned our lights on and broke away from the tanker to the right. "Approach, Eagle 501, ready for vectors and descent for landing" was my call. "Roger 501, continue right to 345, descend and maintain 5000 feet" was the response from Approach. I echoed, "345 and descending to 5000 for 501".

Dog seemed energized. He was ready for this. I'm not sure exactly what had changed but Dog had a fresh perspective to conquer what had been a great challenge this night. The mechanics of the approach were as before. Everything worked well. The approach controller was very supportive and calm. Everyone knew that while we had fuel to do this many more times, Dog may not be up for the task of multiple approaches. The LSO called, "Eagle 501, ¾ of a mile, call the ball". I replied, "Eagle 501, Intruder ball, 6.0". "Roger ball, 26 knots down the angle" came from the LSO. It was time for Dog to perform magic. The whole airwing watched when an aircraft was having a hard time coming aboard. Everyone except the tanker behind us had already landed. Everyone knew of the trouble we had that night. Dog flew the approach like he had when he first got to the Midway. It was a beautiful pass to an OK 3 wire.

Grabbing the wire was a surprise. We had gone around so many times that night that we wondered if it would happen again. Dog applied full power as always. We were thrown forward in our harness. The YS came out in front and told Dog to pull the power back to idle. Power back, lights off, I made a circular motion with my flashlight indicating to our maintenance team that our aircraft was UP and ready to fly again.

Dog raised the tailhook as directed by the YS and added a bit of power to get out of the landing area. I started folding our wings to enable us to park closer to the other aircraft on the flight deck. We taxied up near the aircraft parked near the catapults and started to shut down the electronics and engines of the jet. I looked over at Dog and gave him a slap, "Well done!" He was beaming with pride after flying that last pass and returning from despair. A normal flight from Midway in an Intruder lasted 2.2 hours. This flight was 4.0 hours. The LSOs came around and debriefed Dog on his passes. We hoped that Dog could get out of his slump and back on top with his landing grades.

Sunday 1 November I had five more flights with Dog in the two days following our night of bolters. Three of the flights were at night and Dog remained challenged and colorful with his night approaches to landing. When Dog joined the squadron, he consistently nailed every night pass but his approaches turned into unpredictable, three-dimensional perturbations prior to landing. I tried to help with comments or observations. They didn't seem to make a difference. With each night approach, I became more anxious about his landings. I felt powerless to help and thought a more experienced BN might help him get back on track. After another challenging night approach and landing, I approached the OPSO and requested a crew change. This was uncommon but encouraged if crews didn't feel comfortable with one another. Dog was then crewed with a more seasoned BN, Phil, who had FRS instructor experience. This appeared to help as his landing grades began to improve. I loved flying with Dog but crewing him with Phil was the right move at the time.

Thursday 19 November USSR Intelligence collection ships frequently sailed near our aircraft carriers. They shadowed us for weeks, matching our course and speed while maintaining a position 2 - 5 nautical miles away. One of the Intel collectors had left the Midway battle group and was headed home when the admiral's staff decided to conduct a WASEX with the Soviet ship. We were off the coast of Sri Lanka enroute to the Arabian Sea. Our simulated attack launched at 420pm. The attack force was two A-6E Intruders and an EA-6B Prowler to help with Electronic Surveillance Measures (ESM). The EA-6B could listen for emissions radiating from the ship. I was fortunate to brief the flight as the mission commander with K9 as my pilot. We had an approximate position for the ship, but it was a many hundreds of nautical miles away from Midway. When looking for something at sea, even a ship, it was always a good idea to have a good approximate position. The ocean is vast and finding anything can be like looking for a needle in a haystack.

I sought out our intelligence officer, Vern. He had an approximate position, but it was many hours old. We briefed the flight based on these coordinates. The primary EA-6B Prowler crew was LT John Carter (pilot), Commander Justin (Noel) Greene (CO of VAQ-136), LT Doug Hora and LT Dave Gibson. As this was a significant mission that the admiral wanted to succeed, a spare EA-6B crew sat in the brief to ensure that one Prowler would fly despite any unforeseen problems that might arise. I don't recall the names of the spare EA-6B crew nor the crew of the other Intruder. The primary Prowler crew names were etched in my memory due to the following events.

The plan was to launch and join up as a flight of four (two attack Intruders, a KA6D inflight refueling aircraft and the EA6B) overhead the ship. Once together, we planned to head out to the coordinates provided by Intel. As the Soviet ship was over 400 nautical miles away, the attack force needed refueling inflight to successfully complete the mission. After 45 minutes into the flight, each of the WASEX aircraft took 3500lbs of fuel from the tanker.

We briefed the following hi-low-hi altitude surprise attack on the Soviet ship. The Intruders would fly with their radars in silent to prevent the Intel collector from knowing that we were coming. The Prowler would lead the formation and listen passively for the electronic signal given off from the navigation radar aboard the ship. Once identified, the Prowler would calculate a position for the ship and dispatch the attack aircraft on their strike profiles at the appropriate distance. Once the attackers were dispatched, the Prowler's mission would be complete and the crew would return to the carrier. The Intruders would turn on their search radars on the low altitude, inbound legs for simulated Harpoon missile attacks. Ideally, this would be the first time that the Soviet ship knew that we were in their vicinity. We would fly inbound to the ship at deconflicted altitudes for photo reconnaissance. Once we had flown past the ship, we would return to the carrier as single aircraft for a night Case III recovery.

Administratively, there were a few more items to cover in the brief. The launch and landing were to be in Electronic Emissions Control (EMCON) conditions (radio and radar silence). We practiced turning off all electronic emissions to diminish an enemy's ability to successfully attack the carrier. The EMCON procedures were not standard carrier procedures but were easy to manage once reviewed and discussed. The aircraft would not emit any electronic signals until well away from the carrier to include radars and radar altimeters. The crews climbed to 200 feet and flew out to 20 miles before climbing to altitude and joining up at the briefed radial and distance from the ship. LT Carter had a question about the EMCON procedures, so we discussed how to implement them. The crews understood the plan and we broke up into individual aircraft crews to brief launch specifics, crew coordination and emergency procedures.

The Prowler Ready Room had an adjoining door to our ready room. Our squadrons got along well, so I walked through their ready room enroute to get my flight gear. As I passed by the primary Prowler crew, I noticed that they were still discussing the EMCON procedures. I gave them a nod and told them that I'd see them on the flight deck. I went over to CVIC where the Intelligence staff worked. I explained to the commander in charge, that our mission needed better coordinates to succeed. He understood my request and told me to wait as he disappeared into a back room. A few minutes later he returned with coordinates handwritten in pencil on a small piece of scrap paper. He said, "Here you go. Try these." as he smiled. I thanked him for his help.

Our flight gear and maintenance control were right across the hall from CVIC. That was always convenient. I strapped on my flight gear, checked in with maintenance control and walked upstairs to preflight the jet. All things looked good, so we strapped in and started our engines and waited for our turn to taxi to the catapult. As we waited, we heard that CDR Greene's aircraft was having a maintenance problem. The aircraft was down for flying and the spare aircraft had to take their place. That was disappointing but we had a mission to accomplish, so we launched and commenced the plan.

After taking off and joining up, we flew toward the handwritten coordinates and waited to receive our 3500lbs of fuel from the tanker. The tanker took over as the lead of the flight and cleared us in one at a time to receive our fuel. We flew on the tanker's left wing before receiving fuel, then shift over to his right wing after refueling. Once complete, we took the lead and released the tanker for his return to the carrier. We still had a long distance to fly to get to the coordinates.

As we closed on the ship's position, we waited for a hand signal from the EA6B that they had received an electronic signal from the Soviet ship. Once confirmed, we gave them the lead of the formation with an Intruder on either wing. At the prebriefed distance, the Prowler dispatched us, we executed a hard left turn away and descended to initiate our attack profile. I dropped a waypoint at the distance designated by the Prowler crew then used it for our attack navigation. We descended to 200 feet above the water and increased our speed to 480KGS as we approached the Soviet ship. When I turned on the radar, there was a ship near the cursors. I refined the targeting, stepped our aircraft into attack mode and prepared for a simulated Harpoon missile launch. K9 received launch symbology on his ADI and pulled his designate trigger on his control stick to complete the launch simulation. We proceeded to the ship, took pictures during the fly by, waved to the Soviet crews and commenced a 180° turn as we climbed away from the water. The sun set and darkness was growing. We had a solo climb to altitude and quiet ride home.

We climbed to an altitude that deconflicted with airline traffic yet conserved fuel for our long, overwater journey home. We anticipated an EMCON recovery as briefed. Strike was broadcasting for everyone to conserve fuel and Midway's TACAN was on. That was puzzling for us. We proceeded to our holding point and waited at our maximum conserve fuel speed while waiting for instructions to commence our approach. We received our approach commence time and K9 set up a racetrack pattern to reach our holding point fix at that time. Our approach and landing were uneventful, we parked the aircraft and proceeded down to maintenance control. The aircraft had flown a great distance over water with no maintenance problems. I thanked the CVIC commander for the coordinates. His help and the Prowler crew made the mission a success.

We returned to our ready room to hear tragic news. CDR Greene and his crew had launched on the next cycle after their jet was repaired. They were supposed to check in with the E2C for their mission but never appeared. The ship had been trying to get in contact with them for the past hour but there had been no response. We hoped that a radio failure was causing them trouble and we would hear from them soon. We debriefed our flight in CVIC and provided our camera with pictures as proof of our success. K9 and I had dinner still thinking about CDR Greene and his crew.

We learned the next day that Ironclaw 606 and her crew were lost at sea without warning. Aircraft were sent around their last known position to look for debris. Nothing was ever found. CDR Greene was to record his 1000th trap aboard an aircraft carrier on that flight. His squadron had a cake waiting for him in the ready room to celebrate. They were in mourning instead. A memorial service was held on the hangar bay for the crew a few days later. Their loss was a big shock for all of us as they were great men. It was surreal that they were all gone without a trace[137].

Monday 21 Dec Iranian forces had attacked Kuwaiti tankers in the past prompting the U.S. government to assist their government by escorting their ships through the Strait of Hormuz. On this night, the U.S. military escorted a Kuwaiti tanker through the Strait in support of Operation Earnest Will[138]. The Iranians had previously used Silkworm missiles launched from their shores to attack the tankers. We knew the location of the missile launch sites. Our Intruders bombers were loaded with four MK20 Rockeye II Cluster bombs to destroy the Silkworm missile sites if they launched toward the convoy. Landing with the bombs limited our landing fuel to 4000lbs. That was less than desired and put more pressure on the pilots to trap on their first pass.

Ed and I were crewed to support the escort. We had many flights together since our first at Cope Thunder in January 1986. I appreciated his enthusiasm and excellent flying skills. Ed was very excited to be carrying live bombs with a significant set of targets. He looked forward to an opportunity to protect the tanker from the Iranian threat. I was nervous as my coordinates for the targets needed to exactly match those for the command-and-control group. If ordered to attack, everyone needed to be 100% certain that we were attacking the designated target as

---

[137] The Ironclaw 606 mishap summary; https://aviation-safety.net/wikibase/57214
[138] Operation Earnest Will existed from July 1987 – September 1988; https://en.wikipedia.org/wiki/Operation_Earnest_Will

there were many options. A mistake on the manual entry of one digit of the latitude and longitude of each target could have serious implications on the success of the mission. We launched at 4pm and made our way north to our holding position in the Gulf of Oman at 35,000 feet. Sunset was at 524pm. Our tasking was to orbit in position, monitor the Strike frequency and be ready to attack in the event of a threat to the convoy. If a Silkworm missile site launched at the convoy, we would attack that site.

Aircraft involved in the operation were Navy E-2C, Hornet, Prowler and Intruder aircraft, Air Force fighters and a KC-10 inflight refueling aircraft. The E-2C served as the on-site battle coordination center. Initially, there was a lot of chatter on the radio but it quieted down once darkness set in and the ships started the transit through the Strait. In the Strait, Navy warships positioned themselves between the Iranian shore and the Kuwaiti tanker. The transit through the Strait of Hormuz was done under the cover of darkness to preclude visually acquired Iranian weapons. Iran possessed F-14 Tomcat aircraft but had significant maintenance and aircrew training issues, so they were not expected to be a threat. Our fighter presence was sufficient to destroy them if they got airborne. As we flew high above the Gulf of Oman, all was quiet on the radio as we patiently awaited news of the convoy's progress. They were not visible to us. Our time on station started out rather uneventful. We flew racetrack patterns of 20nmi in length at maximum conserve fuel while listening intently to the radios to determine if our services were required.

We were scheduled typically for a 2.2-hour mission with a night recovery aboard the carrier. Strike kept track of our fuel state throughout the mission. As the time approached for our departure and recovery aboard Midway, Strike called and asked if we could stay longer on station if we received fuel. We were happy with the idea as we hadn't done anything to grow tired. The whole mission was quite exciting and we were proud to be part of it. Strike gave us a vector (heading and distance) to the tanker and asked us to report it when in sight. We looked toward the heading as Ed brought the aircraft around. Despite the tanker being over 50 miles away, it was immediately obvious to us. It was lit up like a Christmas tree and had a formation
of aircraft on each wing. We reported the tanker in sight and proceeded to join on the port side of the aircraft.

As we came in closer to the tanker, the aircraft size seemed to grow rapidly. The tanker was refueling both Navy and Air Force aircraft. There were eight aircraft buzzing around the tanker like bees around a flower. I had a great sense of patriotic pride as we flew in formation on the wing of the KC-10. The aircraft itself was impressive but on its wings were various aircraft from the USN and Air Force. From this spot, we saw the impressive US government commitment to protecting Kuwait. We waited for our turn to refuel. As the aircraft in front of us moved away from the refueling hose and slid to the starboard side of the KC-10, we were cleared to tank. Ed gingerly slid behind the tanker inflight refueling hose. The sight under the KC-10 at night was spectacular to behold. As it had so many lights, it felt like being below an enormous spaceship. Ed added power and inserted our probe into the refueling basket. The KC-10 basket was like an Intruder's basket, making the tanking familiar. We took about 10 minutes to fill our tanks then slid out to the right and away from the KC-10 and back to our holding position.

The radios remained quiet as we orbited and waited for the signal to act. Strike continued to monitor our fuel state and once again asked if we would stay airborne longer if given fuel. We were still excited, feeling fresh and ready to continue so we agreed to more tanking. Tanking the second time seemed easier as the shock of the lights and size of the KC-10 had diminished. We received our second load of fuel from the tanker and returned to our holding position.

After 10pm, we received word that the ships had passed uneventfully through the Strait and we could return to the carrier. We checked in with Strike, switched over to Approach and executed a straight-in Case III approach to an arrested landing. We had been airborne 5.7 hours. That was a long flight but we both felt great to have been involved in the mission. We didn't drop any ordnance but were prepared to do so. No harm came to the Kuwaiti tanker or any of the support forces.

Thursday 24 December Being aboard an aircraft carrier had its advantages. Despite the challenges of being surrounded by 5000 friends and a long distance from home, we had the ability to draw entertainment from the United Service Organizations (USO). The USS Midway was absolutely blessed to have Bob Hope, Barbara Eden, Connie Stevens and her daughters, Lee Greenwood, Miss America Michelle Royer and the Super Bowl Dancers aboard for a show on Christmas Eve[139]! KO served as Mr. Hope's escort around the ship while my CU Boulder mate and VF-151 pilot, Craw, escorted Ms. Royer. The show was amazing with the hanger deck filled to standing room only with sailors. Lee Greenwood touched a soft spot with many onboard with his song, God Bless the USA[140]. The USO team spent the night aboard Midway to catch up on their sleep to prepare for a show the next day aboard the USS Okinawa in the Persian Gulf. It was a great honor to have the team aboard with us. The next day was also special as the ship served an amazing feast for the crew. Being so far away from home and family on Christmas was challenging but all the excitement helped the time pass easier.

Thursday 31 December The USS Midway anchored off Mombasa, Kenya[141]. Our Eagle aircrew got a group of rooms at a beach resort about 20 miles south of Mombasa. The resort had a great swimming pool where we spent a fair amount of time. We discovered a nightclub in the resort and enjoyed a big New Year's Eve and early morning hours filled with dancing.

Saturday 23 January 1988. This was a crazy and memorable day. We had been on cruise in the Arabian Sea for months. We escorted Kuwaiti Tankers through the Strait of Hormuz to protect them from the threat of Iranian Silkworm missile attacks. During all those months with countless number of aircraft flying, we never sent a single aircraft to our emergency divert field of Masirah Air Base, Oman. This fact impressed the admiral aboard Midway so much that he scheduled an all-hands meeting on the hangar bay for all those who could attend. The admiral started his speech with thanks and congratulations for the hard work and dedication demonstrated by everyone aboard ship. Specifically, the admiral thanked the airwing pilots for not diverting any aircraft to Masirah. These words became a curse once uttered by the admiral.

---

[139] Christmas Eve 1987 aboard USS Midway with Bob Hope; https://www.youtube.com/watch?v=kd0nL9eL9_s
[140] Lee aboard Midway Christmas Eve 1987; https://www.youtube.com/watch?v=9ryAKKFRHmw
[141] From the USS Midway 1987-1988 cruise book; https://www.navysite.de/cruisebooks/cv41-88/053.htm

K9 and I were scheduled for a night flight along with F/A-18 Hornets, an EA-6B Prowler and an E-2C aircraft. After our launch, the ship conducted a General Quarters (GQ) practice emergency. The exercise required the crew to safely report to their battle stations as quickly as possible. The crew secured all passageway hatches to make the ship more survivable in the event of battle damage. The flight deck exercise simulated a crashed aircraft in the landing area with numerous fatalities. The scenario script cited heavy damage to an aircraft requiring it to be picked up and dropped overboard. This task required the mobile, flight deck crane, affectionately named, Tilly. The GQ was scheduled for an hour enabling the ship to train while we were airborne yet needed to conclude in time to land our aircraft.

K9 and I practiced low altitude maneuvering over the dark ocean while conducting simulated combat operations. Once we checked into Strike to receive our Case III recovery instructions, we were directed to conserve as much fuel as possible. We were surprised by these instructions and wondered if another accident had occurred. The instructions from Strike were not SOP. We proceeded to our holding position at the aircraft's maximum conserve airspeed. The air was clear of clouds in holding and we watched the other airwing aircraft below us all flying at maximum fuel conservation airspeeds. It was great to see all these aircraft patiently waiting for landing but this went on for 30 minutes. Approach asked for our fuel states every 10 minutes. We all wondered what could be stalling the recovery.

After 30 minutes, Approach made an infamous radio call, "99 aircraft, your divert is Masirah airbase, bearing 300, change to Masirah Tower, frequency change approved". Again, K9 and I looked at each other with utter surprise. I dialed in the VORTAC frequency for Masirah, as he turned the aircraft to a heading of 300°. We didn't know what had caused this change but would worry about that later. We needed to land at Masirah in 15 minutes. We checked in the Masirah tower and were pleasantly surprised by their clear, English diction and helpful manner. We lined up with the runway and conducted a straight in approach behind other aircraft in the airwing.

We followed a field support truck and taxied over to our parking spot on the ramp. Once he had led us to our parking spot, the support truck rushed off to assist the next aircraft from the airwing. We shut down our jet and climbed down our boarding ladders to unstrap from our flight gear. As we hit the ground, a jeep showed up with British aircrew that welcomed us to Masirah and handed each of us a beer. The Brits directed us to jump in the jeep and they took us to their Officers Club to share this rare opportunity to socialize. Before we knew it, all the aircrew that launched from Midway were in the Club. Each of us had a big smile on our faces.

The E-2C crew explained why we were there. Tilly had been brought into the landing area as part of the GQ exercise, but the engine died and couldn't be restarted. Several Ground Support Equipment (GSE) tow tractors were chained together to combine their horsepower to move Tilly. Unfortunately, the flight deck was worn smooth due to months of flight operations and was slippery due to fuel and oil. The GSE wheels spun as they tried to pull Tilly. Finally, with fuel states getting low and Tilly not moving, the flight was diverted to Masirah. We all enjoyed the hospitality of our new friends. We didn't bring any money with us as the night's entertainment was not planned. The British pilots paid for everything and became our new best friends. At various points, crews peeled off to go to sleep as we weren't sure what the next day would bring. We were escorted to accommodation rooms. K9 and I shared a room with two single beds and were quickly asleep.

Sunday 24 January I woke at 7am to an empty room as K9 was missing. I wandered out of the room and looked around in the daylight. The sky was bright blue without a cloud in sight. The temperature was pleasant despite it being winter. We were in single story accommodation with desert sand all around with no vegetation. There were no other aircrew in sight.

I walked to our jet. K9 was there and had just completed his preflight. It had been refueled overnight. That was a bonus as we didn't need to wait around for a fuel truck. We got a GSE cart over to our aircraft and started the jet. I called tower to get a clearance for our flight. We were given a flight and taxi clearance. No one else was moving on the airfield.

We took off, climbed up to altitude and headed back to our estimated position of Midway. The TACAN came to life and pointed home. We checked in on Strike frequency, entered the daytime pattern and came into the groove on final approach behind the ship. K9 made great corrections along the flight path and landed perfectly with an OK 3 wire. We hopped out of the jet, went down to maintenance control to fill out the paperwork, took off our flight gear then went straight to the aft wardroom for breakfast.

Saturday 13 February K9 and I led a night, five-hour WASEX against the USS Enterprise. Five hours was a long time to be sitting on the ejection seat, but this flight was worth it. The USS Midway was being relieved from our duties in the Arabian Sea by the USS Enterprise. Our crews were looking forward to getting back to Japan and shore duty. Message traffic between Midway and Enterprise directed us to leave the Arabian Sea before the arrival of Enterprise. Midway started heading southeast as Enterprise was just past the southern tip of India sailing northwest. The ships agreed to practice WAS tactics with one other as we approached. The admiral's staff aboard Midway wanted to win the exercise.

Our Intruder, an Intruder wingman and two Intruder tankers planned to launch just before dark to conduct the attack while they were over 800nmi away. We planned to simulate the release of four Harpoon Missiles. The attack profile required a lot of fuel. With a three fuel-tank configuration, an Intruder could record 2.5 hours of flight time with average speeds and no tanking. We were planning a high-speed, low altitude Harpoon attack profile near Enterprise, so we needed more fuel. To fly five hours required another 10,000lbs of fuel per aircraft. The plan had two tankers escorting our two Intruders toward Enterprise. They were to give us as much fuel as they could before turning back to Midway for their recovery. Due to our position on the western side of India, there were no land-based airports if an emergency landing was required. We were flying under Blue-Water Operations which meant that regardless of aircraft emergency troubles, we were going to land the aircraft back aboard Midway. In only the most rare and extreme circumstances, would our aircraft expect to land at a shore-based airport or aboard the Enterprise.

Proper tactics and the element of surprise was critical to execute a successful attack. Enterprise had F-14 Tomcats to defend their carrier from a great distance. The Tomcat's radar had great range, their fuel onboard and efficient turbofan engines allowed positions well away from the carrier. We considered how to infiltrate their defenses. I visited our airwing Hornet pilots who were weapon system experts. I didn't know much about air-to-air radar and wondered how to defeat it. One of the pilots described how the F-14 pulse doppler air-to-air radar worked. In simple and unclassified terms, he noted that the Tomcat crews could easily see us as we approached inbound to the carrier. He noted that if we flew perpendicular to the Tomcat's flight path, we could be invisible on their radar screen for a short period of time. If we could become invisible for thirty seconds, it might give us a chance to change our flight profile and launch our simulated attack on Enterprise. It was a gamble, but it became the foundation of our attack.

We added details to help with the surprise. We flew easterly at an altitude that was typical for airlines, FL370. We flew on the airline airways and squawked an Identification Friend or Foe (IFF) code as if we were an airliner. We disabled our classified IFF response that was reserved for U.S. military aircraft. We obtained the radio and navigation frequencies for the USS Enterprise and listened to them as we approached, hoping to gather valuable intelligence on how our attack was being perceived by the carrier and her airborne crews. We used the Enterprise TACAN to determine our position from the carrier. We didn't radiate our radars until in close and were ready to launch our simulated missiles to avoid detection by the carrier's electronic surveillance teams. Our two Intruders flew as a tight formation to appear as one large aircraft (like an airliner) on radars rather than two smaller aircraft. We flew with one set of navigation and collision avoidance lights to visually appear as one aircraft. Finally, as we reached the point of turning abeam the Tomcat, we turned off our lights and dove for the ocean. The timing of this was critical. If the Tomcats lost sight on radar and we changed altitude in a rapid maneuver, we could confuse them and give us precious time to prosecute our attack. That was the plan we briefed and hoped to execute.

We took off in daylight conditions and quickly formed up the division of aircraft using the aircraft launch strategies learned while onboard the USS Kitty Hawk. We departed Midway discreetly using the number two radio for communication in the division, if needed. It stayed silent most of the mission as the crews knew the brief and executed it according to the plan. We climbed to an intermediate cruise altitude of FL290 as it was easier to tank at that altitude without compromising much fuel. We flew for an hour and burned 6000lbs of fuel before refueling. The attack aircraft needed to successfully refuel airborne to complete the mission. If they couldn't refuel for whatever reason, that aircraft had to abort the mission and return to Midway. Both attack aircraft refueled without a problem so the division of four aircraft proceeded as planned. After another hour, the attack aircraft refueled again with the tankers giving us as much as they could. Once the tankers had given us their fuel, they departed and the attack aircraft proceeded as a flight of two.

K9 and I were in the lead aircraft with our navigation lights illuminated but our anti-collision light turned off. Our navigation lights helped our wingman fly close formation on us despite the ever-darkening sky. Our wingman had his anti-collision light on. We tuned up the USS Enterprise Strike frequency used to communicate with the Tomcat crews. I turned off the radio squelch as the initial signal was barely audible. We heard the typical chit-chat of carrier operations. There was no indication that our approach had been noticed. As time passed and the radio signal grew stronger, I turned on the radio squelch to get rid of the perpetual hiss of static.

A moment arrived when the E-2C from the VAW-117 Wallbangers noticed us. Banger Five radioed the Tomcat CAP from the VF-213 Black Lions, "Lion One, you have airline traffic to the northwest, flight level 370". Lion One responded, "Lion One, roger, looking". The serious part of the game had begun. We were anxiously waiting for the Lion aircraft to respond to determine our next move. "Banger Five, we have radar contact with the airline traffic. We will monitor" came from Lion One. The E-2C responded, "Banger 5, copies". We estimated our range to the Tomcat to be about 100nmi.

The commercial airway ran adjacent to the Enterprise with a closest approach at 80nmi. We needed to turn inbound to simulate the release of our Harpoon missiles. So far, the Lion aircraft saw us as an airliner and not a threat. That was good so we stayed on course. The next few minutes seemed like an eternity as we waited for more words from Lion One. We turned off all our cockpit lights in preparation for the engagement that was about to come. As we came closer, Banger Five became more inquisitive, "Lion One, can you investigate that airline traffic at flight Level 370?" 'Lion One, affirmative", came from the Tomcat. The game was about to get much more interesting as the Tomcat was coming to have a look at us. We were to be intercepted but we needed to get closer to Enterprise for the attack. We turned to the left and added full military power to close the distance as fast as possible. As we were at FL370, we weren't using much more fuel than we would at maximum range speed. "Lion One, the airline traffic has changed course and is headed inbound to Enterprise" came from Banger 5. "Lion One, copy. Outbound" came the response with more resolve in RIO's voice. The Tomcat crew were changing into their superman costumes.

Our wingman had been tucked up close to help simulate an airliner, but that game was over. We kissed him off, turned off all our lights and broke into a steep, descending left turn at our maximum Gs. Our wingman broke to the right as he turned off his lights. K9 turned perpendicular to the inbound course to Enterprise in the hope to drop off the Tomcat's radar screen. After about 10 seconds, he reversed continuing the steep, descending turn back toward Enterprise. We were screaming down toward the black sea passing 15,000 feet when we looked up to see the lights of the Tomcat well above us. Lion One was in a turn looking down into the pitch-black ocean toward us. "Banger Five, Lion One, we're not sure what we have here. We're blind[142]". There was no chance for Lion One to see our dark aircraft against that dark sea. The moon was in its waning crescent phase and had not yet risen over the horizon. Lion One didn't have any ambient light to help acquire us visually.

As Lion One circled overhead, K9 commenced a hard 360° turn to stay below them and eliminate their radar acquisition. This was a fun cat

---

[142] This meant that Lion One had lost radar and could not obtain visual contact with us.

and mouse game that the Tomcat didn't anticipate. If they could see us, they could kill us, but they couldn't see us. After the 360° of turn, we proceeded inbound toward Enterprise at full power and an altitude at 5000 feet. I turned on our ground mapping radar, acquired then designated the USS Enterprise with my cursors, stepped the mission computer into attack for the missile launch. K9 committed to the release and our aircraft performed a simulated release of 2 Harpoons. We broadcasted on the Enterprise Strike frequency, "Bulldog One, Bulldog Two" notifying the ship that two simulated Harpoon firings had just been conducted against them. Ten seconds later, we heard, "Bulldog Three, Bulldog Four" from our wingman. K9 turned on our lights then snapped on a climbing left turn while looking out to ensure the Tomcat was not nearby.

Once back on a northwest heading back to Midway, K9 rolled out of the turn but kept the power at military and speed at 290KIAS. Around 25000 feet we reached M 0.7 and held that until we reached our Bingo designated maximum altitude of 43,000 feet. We had a long distance to get back to Midway and needed to conserve as much fuel as possible. Upon reaching the cruise altitude, we checked the airspeed (400 KGS), distance to go (700 nmi) and fuel required to get there. The Bingo table indicated that we needed 4300lbs to get home and we preferred 6000lbs on the ball for a total of 10300lbs. We were going to be 1000lbs short but would manage.

We had a chance on the flight home to discuss and relive the attack. K9 had surprised me with his improvisations during the engagement with the Tomcat. His 360° turn below the fighter was the right thing to do but we hadn't briefed anything like that. No doubt the Lion One crew were scratching their heads wondering about the events. We guessed the post-flight conversation with Bangar Five was an interesting one. We delayed our descent and requested priority routing from Strike when we checked in on their frequency. They were happy to comply and we recovered with 5.1klbs of fuel. Mission accomplished and the admiral's staff were pleased to have won the WASEX.

As the USS Enterprise was travelling west to relieve us in the Arabian Sea, the USS Midway was sailing east back to Japan. As escorting the Kuwaiti tankers presented the possibility for combat any day, our squadron prepared a turnover folder for the oncoming Intruder squadron, the VA-95 Green Lizards. The folder contained an intelligence summary with data that had been collected over months during our cruise. As it was the best, most concise information on the Iranian threat, we hoped it would help the Lizards quickly prepare for combat. The classified document was properly wrapped and delivered to the USS Enterprise in late February. On 18 April, VA-95 flew in Operation Praying Mantis resulting in the squadron sinking an Iranian ship and boat and damaging others. VA-115 had been ready for combat but things didn't escalate until VA-95 had the duty. Maintaining a daily capability to conduct combat operations took focus, discipline and a great deal of hard work.

Midway made her way to Thailand in late February for an in-port period in Pattaya Beach then sailed toward the Philippines and South Korea for Team Spirit.

Wednesday 6 April KO and I flew off Midway and landed at Atsugi marking the end of our Persian Gulf cruise. The cruise had lasted a bit over 5.5 months. We were happy to be home. It was a nice feeling to have our own BOQ rooms after living together at sea for a long time. The first week after the return from a long cruise was quiet. The airwing didn't have much flying money, so it was a time for the married guys to renew relationships with their families and everyone to plan for the next set of activities.

Friday 13 May VA-115, VA-185 and VAQ-136 had a detachment planned for Osan AFB in mid-June. I was managing the detachment and prior coordination with the Air Force was required. Pilot Larry and I flew to Osan in an Intruder for a range scheduling conference to secure bombing range and low-level times. As we sat in the room full of Air Force pilots, they took ALL the low-level and target time slots. We couldn't get anything to support our detachment. I raised the BS flag in the meeting explaining who we were, how our training opportunities were extremely limited and what we needed. An Air Force Major in Osan Operations went to bat for us and helped get us low level and target times. We were greatly indebted to her. The Air Force were more generous with night times as most weren't equipped to fly low levels at night and preferred day bombing. With these times, we had the foundation that justified moving VA-115, VA-185 and VAQ-136 to Osan for a week.

Friday 10 June CDR 'Polecat' Polatty relieved CDR Cash as the Commanding Officer of VA-115. Our new XO pilot was CDR T. Toms[143].

Sunday 12 June New pilot, Joe and I shared our first flight together. He was a stocky, athletic character with a calm and collected disposition. Joe had the looks of a Hollywood movie star and loved being overseas with the Eagles.

Thursday 16 June A new pilot with prior Army helicopter experience, Gunny and I flew over to Osan for the start of our detachment. He was a solid stick with a jovial demeanor. With prior helicopter experience, Gunny loved flying the Intruder. We were quite excited as the training would provide unique opportunities. Our bombing range and low-level times worked well and we enjoyed a great training experience over a two-week period.

Friday 1 July As a way of saying thanks to the Air Force officers that assisted us, we threw a party at their beautiful and new officers club. We purchased eight, 55-gallon trash cans and filled them with beer and ice. They were placed on the outdoor patio area at noon, so the beer had plenty of time to chill. We told our Air Force friends about the 4pm commencement and ordered food for their arrival. It was a warm yet pleasant afternoon with jets returning within sight of the club balcony. As the Air Force pilots and staff started to arrive, they were surprised by our generosity with food and drink. We explained our appreciation for their support for our detachment. The club became very busy as our party drew a crowd. The beer flowed and everyone had a great time. Some HH-3E Jolly Green Giant helicopter[144] pilots were happy to join the party.

As the evening continued, an Air Force pilot suggested that we do 'carrier quals'. The long tables were set up and wetted down with beer.

---

[143] CDR Toms was the Eagle CO during Desert Storm and later served as the Commander of Attack Wing, U.S. Pacific Fleet.
[144] These are big aircraft; https://www.nationalmuseum.af.mil/Visit/Museum-Exhibits/Fact-Sheets/Display/Article/196060/sikorsky-hh-3e-jolly-green-giant/

We needed a towel or tablecloth to serve as the arresting gear cable like those aboard aircraft carriers. An Air Force flight surgeon and I went into a back storeroom and found a large white tablecloth to use. We rolled it into a cable and positioned ourselves on either side of the approach end of the table. A 'participant' ran at the long end of the table and jumped belly first while catching the 'cable' with his flight boots/shoes. If the participant missed the cable, he slid down the table falling off the other end. The crowd made themselves very vocal if that occurred.

The quals started with a Navy pilot showing his hosts how it should be done. He landed successfully and 'caught the wire'. Air Force pilots lined up to have their turn at catching the wire. You could tell that many of the pilots had done this before as they were quite good at it. Naturally, as each pilot came running toward the table there was loud vocal encouragement from the crowd. It was noisy, but all in good fun. Everyone was having the time of their lives.

The flight surgeon and I turned over the cable holding duty to other pilots and we watched on as the merriment continued. A Jolly Green Giant pilot decided to have a try. This guy was big, probably 6'4" and 230+lbs. He was a giant of a man. As he started his run, it was like watching a big train moving slowly at first then building up a big head of steam. As he neared the table, he jumped up high into the air rather than down the table. My flight surgeon friend and I gasped. What goes up, must come down. Down he came with a loud crash, smashing the table with his landing. The crowd went wild with screaming, laughter and whistling. It was a sight to behold. Fortunately, our giant of a friend was not injured. The next day I saw one of the Air Force officers. He commented, "That was the best party we have ever had on this base. We had not christened that new O' Club until last night."

Before leaving this Osan detachment story from June 1988, it is worth adding this bonus. While in Germany in December 1990 for a rare international trip with VX-5, an amazing meeting occurred. Our test team visited an officer's club at a military base. I was at the bar drinking a beer when a conversation started with an Air Force flight surgeon. We discussed the doctor's duty locations during his career. He had served at Osan AFB in 1988. I mentioned that our squadrons from Airwing Five had a weapons detachment to Osan in June 1988. The doctor mentioned a party at the new officer's club on Osan that was hosted by navy flyers. He noted that it was the best party they had ever had at Osan. I explained that I was the coordinator of the party. He became very excited to relive the memory. He recalled the carrier qualifications when pilots ran and jumped stomach first onto a table wetted down with beer trying to catch a simulated wire with their feet. I verified his memory recalling that an Air Force guy and I went to a linen cupboard and found a white tablecloth that we rolled into the cable. He suddenly looked at me differently and with a big smile on his face said, "I was the guy that found that tablecloth with you!" We burst into laughter.

What were the chances of this meeting in Germany in December 1990 after the party in South Korea in June 1988? We laughed recalling the Jolly Green Giant pilot who smashed the table and the wild crowd that responded. The doctor noted, "It was the best party we ever had at that club". It was gratifying to hear those words.

Thursday 7-8 July Gunny and I had an overnight cross country to Osan AFB. As things turned out, it was our last flight together as I transferred back to the States in early September. Gunny was a great pilot and I miss flying with him[145].

August 1988 The Eagles had a hail and farewell party for a couple of officers (including myself). As accustomed for departing Eagle officers, I gave a speech. I thanked the Eagle family for years of amazing memories. I thanked the wives for all their creativity and energy in making cookies, sending cards of encouragement and presents to the bachelors as well as their husbands while we were on cruise. I mentioned all the parties and good times.
I mentioned how the tour had provided a deep low (the loss of the crew of Ironclaw 606) and many highs. I closed with a commonly heard phrase from H-Mac, "If you're not having fun, you're not doing it right". On that note, it was time to move onto the next adventure.

---

[145] Guy 'Gunny' Gribble was an amazing man that served his country and American Airlines with distinction; https://www.dignitymemorial.com/obituaries/arlington-tx/guy-gribble-9870023

# Chapter 8

## Military Flight Instructing at NAS Whidbey Island

*Murph had performed so well at the boat then again at Cecil with a challenging approach on a dark night. I could see his surprise yet pride in the performance.*

After living nearly three years overseas, I felt a need to reconnect with the USA. Upon arrival at NAS Whidbey Island, I had a wonderful feeling of being back home. The scenery was stunning and everything was familiar. A lot had happened since leaving VA-128.

Tuesday 18 October Checking into VA-128 required visiting various departments around the squadron. All instructors had ground jobs to perform while they were not flying. Having just organized a three-squadron tactics development detachment at Osan AFB, I imagined a job fitting to this accomplishment. I reported to the OPSO who told me that my new job would be.... drum roll inserted here.... the Coffee Mess Officer (CMO) for the 180-man wardroom! I'd been the CMO at VA-115 and hoped to have graduated from this duty but just needed to make the best of it.

I began my Instructor Under Training (IUT) syllabus within a few days after arrival. We learned the theory of how people learn and how to teach effectively. I had nine Intruder and two TC-4C Bombardier training flights in the instructor syllabus. Six of the nine Intruder flight were IR-342, IR-344 and IR-346 day/night low level flights with visits to Boardman bombing range. The TC-4C flights were a radar bombing flight to Spokane and a mine laying flight in Admiralty Bay across from Port Townsend, WA.

17 November I had an Intruder NATOPS Instructor check-ride in the WST with my first VA-115 pilot, Waldo. He was an IP at VA-128. USMC IP, Ray, tortured us with emergencies from his console. The session proceeded smoothly. It was great having Waldo as my pilot for the simulator as our crew coordination was already refined from flying as Eagles.

Saturday 7 January 1989 An Intruder FRP, Mark, asked me to attend his wedding. I was naturally honored but a bit surprised as I hadn't flown with any students and was a new instructor on the scene. Maybe he liked my performance as the CMO. The wedding and celebration afterwards were held at his apartment. There were other students in attendance. As things turned out, Mark became a decorated combat pilot with the VA-115 Eagles, a test pilot, an astronaut (along with his twin brother Scott), later married Gabriel Giffords (GG)[146] and is now a U.S. Senator[147]. Mark is living proof that with hard work and skill, you never know what life might have in store for you! Life is a box of chocolates…

Friday 13 January My first scheduled flight with a student was combined with a cross country. My student, Rich, was a confident young man succeeding in flying the Intruder. He was ready for his first day, low-level flight to Boardman for bombing practice. He also needed instrument flight-time, so we considered the options of how to combine the priorities. We applied for a cross-country including the low level, Boardman bombing then instrument flight time enroute to Hill Air Force Base, Utah. The request was approved and we looked forward to the flight.

We took off from Whidbey and once east of the Cascade Mountains descended toward the flat farmland of eastern Washington state for our low level at 360KGS and 200 feet AGL. Rich was doing a great job of flying the jet and our crew coordination was good. As we approached Boardman, I made our standard radio call to check in and proceeded into the bombing pattern. We finished our bombing and commenced as climb toward Hill AFB. After landing at Hill AFB, we refueled the aircraft and proceeded to the BOQ for accommodation.

We shared an instrument return to NAS Whidbey on Sunday. We landed with a total of 4.7 hours of cross-country flight time. Rich and the squadron operations department were happy that we completed the flights and furthered his training.

---

[146] GG; https://www.youtube.com/watch?v=1_via73LqPs and https://www.youtube.com/watch?v=M6TMSQWI9m4
[147] Mark Kelly; American Hero. https://www.kelly.senate.gov

Sunday 29 January VA-115 BN Jay Cook died during a night, hot refueling accident at NAS Cubi Point. Fuel was sucked down the right engine and ignited causing a fireball. Jay ejected but was out of a safe ejection envelope[148]. Jay arrived at the Eagles just as I was leaving in September 1988. This was a very sad time for the squadron.

Thursday 9 February. Our OPSO, IP Pat[149] and I flew down to El Centro, CA for a Visual Weapons Detachment that lasted until 23 February. We taught the students how to safely complete day and night 40° dives, level laydowns and popup bombing along with ACM. The 40° dive bombing pattern required dive after dive followed by 4 - 6 G pullups. I had a flight with FRP Rich completing a dozen dives in the afternoon. It was great to see him gaining confidence and mastery of the Intruder.

One night, Rich and a few other students got up on stage at a bar and played rock and roll songs. I was impressed and started talking to them about their music. Apparently, the class had talented musicians and they had been practicing. This was the start of something great that blossomed in time.

Thursday 2 March FCLPs commenced for those students ready for their carrier qualification in the Intruder. I was teamed up with FRP Murph, an easy going, funny guy with brown hair and a bushy mustache[150]. Murph and I hadn't flown together before, so I wasn't sure what to expect as we did a day bounce, hot refuel and night bounce. He was solid from the beginning. That made my job easy with only a few comments required as he flew the approaches. In the weeks that followed, we completed six simulator sessions and sixteen day and night bounce periods. These prepared us well for Murph's first Intruder carrier landings aboard the USS America. As an instructor, I was generally quiet allowing him to concentrate yet provided advice when needed. In general, Murph performed well requiring little commentary or assistance.

---

[148] A summary of the mishap can be found here; https://aviation-safety.net/wikibase/186149

[149] Pat had been my pilot when landing aboard a carrier for the first time as a student.

[150] Murph reminded me of a surfer dude. He was the drummer for the student band.

Monday 20 March Murph and I had a FCLP warmup flight to get ready for carrier landings. He flew well and was ready for the boat.

Tuesday 21 March Murph and I flew from Cecil, landed aboard America and completed 6 daytime, arrested landings. His passes were solid and presented no surprises. We flew back to Cecil to wait for our night flight. We needed six, night arrested landings to finish his qualification. We flew out to America and received vectors for a Case III approach to landing. Our initial approach proceeded well with Murph flying the needles. As we flew inside of two miles, Murph started wandering back and forth across the extended centerline while doing a good job of maintaining on speed and glideslope. Inside of a mile, he was overshooting his lineup corrections. I reported his vertical velocity about every 2 seconds to make sure he didn't develop an excessive rate of descent. Other advice helped smooth out his control inputs. Murph appreciated the comments as they were timely. As he made lineup corrections, the upper wing spoilers decreased the wing's lift and increased drag resulting in a higher rate of descent and a touchdown below glideslope. We trapped but it was not ideal.

We hoped the first trap was a warm-up and Murph would get better on the next approach. Our first night catapult together was uneventful. It was Murph's first night catapult but his experience with the simulators at Whidbey prepared him for it. After launching off the deck, we entered a Case III pattern following CATCC's instructions. The next approach was nearly identical to the first repeating the problems of holding centerline and landing below glideslope. The LSO commented over the radio for Murph to keep a higher glideslope in close. We acknowledged the call. Murph started getting nervous. He wanted to do well but wasn't sure why we were getting low in close. We discussed the need to maintain alignment with the extended centerline of the landing area. Murph agreed to spend more time focused on holding centerline.

We were catapulted back into the dark night and followed CATCC directions again. Everything worked like clockwork delivering us on final behind the carrier. Murph spent more time trying to hold centerline, but the wandering remained and we ended up low on final. We launched again with similar results on the fourth trap. Murph wasn't getting any better at holding centerline or glideslope in close. The LSOs wanted to discuss our first four approaches and landings. We were directed to parking, shut down the jet and were escorted to our ready room.

Murph was anxious as we waited for the LSOs to finish the night traps and debrief our passes. We saw our passes replayed on the TV screen in the ready room. We could see that chasing lineup was a problem but were uncertain how to fix it. The LSOs confirmed our perceptions noting the lack of centerline control was resulting in spoiler use and below glideslope landings. If Murph could fix the centerline tracking problem, his approaches would improve. We all understood the problem but needed a commonsense solution.

One of the LSOs provided a PEARL of wisdom that saved the day. He asked, "How far away from the centerline of the aircraft is the center of the pilots ejection seat?" Our answer was about 2 feet. The LSO continued, "If you land with the centerline right between the pilot's legs, is anyone going to be upset?' The answer was no, as landing within 2 feet of centerline was fine with everything else that was happening. The LSO then directed, "Murph, as you transition from the needles to the visual approach, place the runway centerline between your legs and keep it there with small corrections all the way to touchdown". That was GOLD! We slept on the carrier and had time to digest the idea.

Wednesday 22 March Murph and I were given three daytime cats and traps to warm-up and try the new strategy of holding centerline between his legs. Murph landed with the centerline between his legs on each pass. Murph's confidence soared! He had a solid strategy to solve the night landing problem. We gave up our jet in a hot switch and went below to have lunch. We discussed the day landings and Murph felt much better.

After darkness set in, we catapulted off the deck, entered the Cat III pattern and headed inbound to the carrier. Murph flew the needles well then transitioned to a visual approach as we flew inside of 2 miles. He kept the runway centerline between his legs and we landed with an OK 3 wire. We were very excited. Why didn't the LSOs tell all the pilots to keep the runway centerline between their legs at the beginning of FCLPs? It would have saved a lot of aggravation and anxiety with everyone[151].

We catapulted off the deck and repeated the previous pass. Murph kept the centerline between his legs and the glideslope became easy to manage. We trapped with another OK 3 pass. The LSO called on the radio, "Good passes, Murph. You're a qual. We need you to fly to Cecil for the night. Are you ok with that?" I looked at Murph and got a thumbs up. "Copy, Cecil is good for us", I responded. We were launched and started a climb up to 16,500 while making a turn toward NAS Cecil. I checked the ATIS to find that Cecil had a 200-foot ceiling and ½ nautical mile visibility due to fog. An Intruder with two fully NATOPS qualified crew members could fly down to 100 feet AGL and ¼ nautical mile visibility but Murph wasn't NATOPS qualified yet. We contacted Cecil Approach and requested the PAR. Murph had great night carrier landings and now faced a test of his instrument flying down to aircraft minimums.

The Cecil PAR controller did a great job of giving Murph appropriate calls to direct our approach. As we descended below 500 feet AGL, we were in a blanket of white fog reflected by our landing light. Our descent rate was 600 FPM so we had 30 seconds until our Decision Height (DH). If we didn't break out of the fog by 200 feet AGL, Murph had to add full power and climb away from the ground to recommence another approach. Four hundred feet passed, three hundred, two hundred fifty then two hundred. Murph started to add power as we broke out with the runway directly in front of us with sufficient visibility to safely land. He pulled the power back and set the aircraft down on the runway. Murph had performed so well at the boat then again at Cecil with a challenging approach on a dark night. I could see his surprise yet pride in the performance. I was very impressed with Murph's flying and told him so. We celebrated the next night as Murph and all the other FRPs fully qualified at the carrier.

In talking to Murph, he mentioned that the students who performed at the bar in El Centro were still playing music. I expressed interest and they invited me to a rehearsal at FRP Reggie's house on the

---

[151] Years later, I used the strategy to help my civilian flight students with their crosswind landings.

beach[152]. Reggie and Rich played guitar and sang vocals. Murph played drums. One of the Marine FRBN's played bass guitar. I showed up with sandwiches and drinks. They were impressed and joked that I could be their manager. Their music was great and I suggested that we book them to play the Whidbey Island Officers Club. They were surprised yet thought it might be fun. They needed time to develop a couple sets of music so we planned for a date in May. The music they wanted to play was spirited, classic rock and roll that would inspire everyone to dance.

I contacted the Officer's Club manager to discuss the feasibility of booking the student's band, The Golden Retrievers. He was initially hesitant imagining that a band made up of aircrew couldn't be any good. I assured him that they were good and weren't asking for money. They would play for free! The manager couldn't pass up that deal, so we booked a date. I had flyers made up to advertise the event. We posted the flyers at the Officers Club, at VA-128 and around the base. Everybody was wondering who were 'The Golden Retrievers'.

Monday 1 May Murph and I were scheduled to complete his first day/night low level flights on consecutive days. The navigation route for both flights was the IR-342 ending at the Boardman bombing range. The day flight allowed for the visual low-level, VR-1355, back to Whidbey Island if the weather permitted it. We used the onboard radar and SRTC mode to fly the IR-342 at 1000 feet AGL and 360 KGS.

The day flight was Murph's first time flying the aircraft using SRTC. The weather at the start of the IR-342 was good allowing Murph to see the terrain around the aircraft as we flew along. Most of his time was spent focused on his ADI display as it provided a window to the world on night flights or if we ran into poor weather. Despite this, it was always fun and comforting to look out the window and take in the mountains and scenery outside the aircraft. It was a thrill to see how the Intruder radar system could nestle us into the mountain valleys.

---

[152] Reggie became a Desert Storm decorated VA-115 Intruder pilot and continued sharing his musical skills. He was a legend and a great guy; https://www.legacy.com/us/obituaries/gastongazette/name/reggie-carpenter-obituary?id=18251747

We ran into clouds that engulfed the aircraft for about 10 minutes. This was good for Murph to experience as it didn't change how the system performed or our crew coordination. The turbulence from the clouds bounced us around giving Murph experience in compensating for real-world conditions that weren't modeled well in the simulator. We broke out of the clouds as we neared the Boardman bombing range. We checked in with them over the radio and were cleared to drop our single MK76 bomb. I provided precise steering to a weapon release point on Murph's ADI. He squeezed and held the commit trigger on the control stick until our bomb came off. We made our left turn away from the target area and checked out of the range with Boardman tower.

The weather looked clear to the northwest allowing us to fly the VR-1355. I gave Murph a heading of 271ºM. Our minimum altitude on the 1355 per SOP was 200 feet AGL but we remained higher to avoid crop dusting aircraft and power lines until we hit Point B. Once there, I gave Murph a new heading of 315ºM, cycled steering to Point C and verified the bearing and distance.

The terrain at the beginning of the route to point C was rolling hills covered with evergreens. It was quite beautiful but there weren't any valleys initially. About four minutes into the leg, the northern portions of the Klickitat River valley were opening to our left. I pointed it out to Murph and he entered the valley. We stayed near 3500 feet MSL while the adjacent terrain was nearly 7000 feet MSL. It was an exciting feeling to be flying in a valley and looking up at the towering mountains on either side. Unfortunately, clouds ahead of us required deviations to remain clear. We departed the low level and contacted Seattle Center to obtain an IFR clearance back to Whidbey Island.

Tuesday 2 May Murph and I flew the night portion of the IR-342 logging 2.5 hours of flight time. We briefed the flight as a repeat of the day flight without the VR-1355 visual low level due to darkness. Flying the route the day before made everything familiar and simple. That was a good SOP as flying low level in the mountains at night is not a natural thing. It is good to be simple, especially for a new student. We released our single MK-76 bomb, departed the Boardman bombing range VFR and picked up an IFR flight clearance to Whidbey Island returning at high altitude. Murph passed the indoctrination into the world of Intruder night

155

low levels and was impressed. I was pleased to have completed a day/night low level with a student.

The Friday afternoon finally arrived for the Golden Retrievers gig at the O' Club. As Whidbey was located so far away from major cities, social opportunities were rather limited and the club was a focal point. The O' Club started filling with patrons around 4pm. The club provided free hors d'oeuvres for patrons starting around 5pm. We decided to have the band start playing music after the Club was hopping. The band started setting up as patrons were arriving at the club. Around 530pm, Reggie got on the microphone and thanked the O' Club for allowing them to play. A small group of semi-interested patrons stood around the perimeter of the large dance floor, doubting that the group of aircrew had anything to offer. The Retrievers started playing and everyone started looking at each other with big smiles on their faces. The band was very good and everyone could see and hear that!

It didn't take long before the dance floor was full of people loving the music as they danced. The band completed a 45-minute set of music and took a break. They were all excited by the crowd's response. The O' Club manager was happy as the place was buzzing in a new, positive way. The band started another set of music and the dance floor filled quickly. As a few patrons had been drinking since 4pm, they grabbed dining room chairs that had wheels and started chair dancing (rolling) all over the large floor. There were chairs going everywhere, people dancing between them and the band just kept pumping out great music. When the second set was over, the energy was really pumping and everyone was having a great time. Reggie announced that the band was done playing but the crowd wouldn't hear of it. He explained that they only had two sets of music. The crowd insisted that they replay their first set. The band was surprised but happy to accommodate their new fans. They took a break and replayed the first set of music. The dance floor was full when the band finished their third set but they were exhausted. We promised that they would return to the Club soon. It was a wonderful night of clean fun for a bunch of great Americans.

Saturday 20 May Alex Zuyev defected in a Soviet MIG-29 to Trabzon, Turkey and asked for asylum to the United States. I don't recall hearing about this event when it occurred. In time, we heard that Alex was sharing details of the aircraft and Soviet tactics. I never imagined that it would touch my life in a personal way, but it did in 1991 when he visited VX-5 in China Lake, CA.

While instructing at VA-128, news arrived of my selection for Class 97 at USNTPS with follow-on orders to VX-5. I was to report to PAX[153] in June 1989. Russ was one of my favorite VA-128 instructors and a well-respected aviator with VX-5 experience. He told me, "Don't get lost in the acquisition process. Come back to the fleet". I nodded in agreement. I didn't really know what he meant by the comment but learned in time. I was eventually swallowed up in the acquisition system despite his caution.

I was going to miss the staff and students at VA-128. The Intruder community was a focal point for carrier aviation and the FRS was the focal point for the community. There were so many amazing individuals working there and it functioned like a well-oiled machine despite the perpetual challenges. I hoped to see them all again[154].

---

[153] Most flight test of Navy projects originates at PAX; it is located a 1.5-hour drive southeast of Washington, D.C. https://ndw.cnic.navy.mil/Installations/NAS-Patuxent-River/
[154] I had one short trip to VA-128 in 1991 to discuss the tactical use of Night Vision Goggles (NVG).

# Chapter 9

## U.S. Naval Test Pilot School[155] at NAS Patuxent River (PAX)

*Preparation, knowledge, and discipline can deal with any form of danger; that danger confronted properly is not something a man must fear[156].*

My first application to attend USNTPS occurred during my VA-115 tour. I had flown with test pilot, Axel, as a student at VA-128 then with my CO, Dusty and OPSO, Smiley, while with VA-115. I'd met test pilot Jeffrey while aboard the USS Kitty Hawk. They all had impressive piloting skills and knowledge. I chatted with an admiral's staff member, Dan Bursch[157], while aboard Midway. He was a USNTPS NFO graduate and inspiring. Dan was later selected by NASA and served as a Mission Specialist astronaut. Over time, I met other aviators that had applied to or been accepted for TPS. All of them inspired me to apply.

USNTPS held a selection board every six months and were looking for twenty-five aviators with engineering, science and/or mathematics backgrounds. Others could apply and be selected but the course was easier for those with a technical foundation. My engineering background helped me immensely to understand the USNTPS syllabus and the field of experimental flight test. Pilots without a technical background could be an excellent test pilot. Test pilots may not understand the theoretical content of the design or why a particular test is important but are able to deliver the aircraft superbly to a particular test point. A test point can be defined in many ways including a specific altitude, airspeed, Mach number, G-loading, AOB, pitch/roll/yaw values or rates. Flight test teams develop specific tests to stress an aircraft to its design limit (dynamic pressure, airframe stress, etc....) to validate performance. Some test points can be quite challenging to achieve as they require unnatural and possibly uncomfortable maneuvering of the aircraft.

---

[155] It is beyond the scope of this writing to describe all the great attributes of USNTPS. This presentation does an excellent job. Enjoy!
https://aero.und.edu/space/_files/docs/colloquium-series/2020-04-27.pdf
[156] The Hunt for Red October, Tom Clancy, 1984, Naval Institute Press.
[157] Dan Bursch; https://www.youtube.com/watch?v=1zsrNR9otQQ

To illustrate the point, I fast forward to September 1997. A test that illustrates a difficult test point was during a F/A-18F Super Hornet weapon release at a specific Mach number, bank angle, g-loading and altitude. Super Hornet test pilot Broadway and I were crewed together for the release of an AMRAAM missile. The test point required Broadway to set up high then dive steeply to an altitude in a 5G accelerated corkscrew. He seemed clairvoyant as to how to deliver the aircraft to the correct spot in the sky to get the Mach number, altitude, airspeed and G loading to align. I watched him do it repeatedly from the back seat of that F/A-18F and was impressed.

How do you train pilots to have such skill? Pilots came to TPS with proven skill from their fleet performance and were taught flight test technique. As an NFO, our Airborne Systems course at TPS combined aircraft flight test theory along with testing weapons and systems like radars, passive electromagnetic radiation detection sensors including infrared, computers, software, crew interface devices, night vision goggles and others. Preparing for TPS started with an individual's background and passion for aviation, the aircraft and weapons they flew and their ability to communicate their understanding to others. There are many opportunities in a squadron for fleet aviators to demonstrate their aircraft knowledge and tactical employment to their fellow aircrew.

I checked into the BOQ at PAX and headed over to the indoor basketball courts on base. I was shooting baskets when Beef arrived and joined me. He was an east coast F-14 RIO also waiting for our TPS class to commence. We played a few games of one-on-one. I was impressed with his style, hustle, humor and demeanor. This meeting was the start of our long friendship.

I found a two-story townhouse about 15 minutes north of the base. The setting was quiet with a big room off the lounge room that served as my study. USNTPS is about reading, studying and countless hours of writing.

USNTPS Class 97 convened with thirty students from the U.S. Navy, Air Force, Army and Marine Corps, a Swiss Air Force jet pilot, Res and an Australian helicopter pilot, Miz. There were fixed wing and

159

helicopter pilots and engineers as well as NFOs and engineers enrolled in the Airborne Systems course. The Commanding Officer was Captain Dusty Rhoades, my first VA-115 CO. The Airborne Systems Head Officer was Smiley, our OPSO while in VA-115. Despite flying with them at VA-115, I never flew with Dusty and only twice with Smiley while enrolled at TPS, both ungraded flights. The Airborne Systems students were Beef (F14 RIO), Weeds (F-14 RIO), Spanky (E2C NFO from USS Midway), Bruce (P-3 NFO), Rich (EA6B NFO), Toad (A-6E BN from VA-115) and Slider (Electrical Engineer). Spanky earned his callsign by resembling the child actor from The Little Rascals comedy.

Before sharing experiences at USNTPS, it is worth stating that I won't burden the reader or myself with trying to convey all the technical details of the school. Firstly, I'd do a poor job trying to explain it all. Secondly, it is far beyond the scope of this writing. Thirdly, many of us didn't understand ALL the math and science presented at TPS and it was considerable in depth and volume. It would be futile to try and relearn it and share it now after all these years. Fortunately, for many test aircrew and engineers, we didn't need to understand all the math and science to do our part and contribute. There is truth in Elton John's Rocket Man song, "And all this science I don't understand. It's just my job five days a week". Our class commemorated this theme with our class patch that is designed to look like a radio station logo; WTPS 97.0 FM. The FM stood for 'F'ing Magic' as that is what makes sophisticated aircraft fly along with many brilliant minds, hard work and money!

The civilian academic instructors at TPS were exceptional. They provided insight and details about aircraft design and testing that was far beyond our imagination, or desire to know at times. They tried to present the material in a most entertaining way to not confuse, depress or irritate us. JJ was our Pitot Statics and Aerodynamics instructor. He was like a game show host for aeronautical knowledge as oozed enthusiasm for his topics. Doc Richards taught mathematics and looked like the Doc from the movie, 'Back to the Future'. He was much less excitable and clearly more brilliant than the movie character.

John taught Performance, had a very dry sense of humor and often we weren't sure if he was joking or telling the truth. He was quietly brilliant but would stay reserved until someone asked a question then unload his knowledge onto our unsuspecting brains. We learned not to ask many questions. Our Communication Systems teacher, Dr. Masters, was at times difficult to follow. He kept talking about frequency folding and aliasing and bandwidth. It appeared to have no relevance to anything ever in our mortal lives. Modern WiFi is founded on the topics of his teaching, but we were too hard-headed at the time to comprehend the material.

Tuesday 11 July 1989 Class 97 had a quick meeting in the conference room at 3pm. We met a couple of our instructors, TC and Joe. They explained that we would have academic classes in the morning and flights in the afternoon for the first six months then swap schedules with the new junior class. If we didn't have an afternoon flight, we were free to manage our time and assignments.

Wednesday 12 July There wasn't any assigned seating in the classroom. I ended up with the best seat in the room. It was in the back, right corner away from the instructor and right in front of F/A-18 Marine Hornet pilot, Smash. He has a wicked sense of humor and passion for the Three Stooges comedies. Smash's passion was more like an obsession. I was fortunate to be at the receiving end of his incessant, Three Stooges monologue during the eleven months at TPS. Smash would translate the instructor's teaching into Three Stooges jokes that he recited in a low voice, audible only to me. It didn't appear that any of our classmates or the instructors even knew it was going on. I sat laughing on the inside but expressionless on the outside. That was a great way to get through the challenging courses at USNTPS!

Thursday 13 July Our class had a math test in the morning to assess our skills before we started tackling the syllabus. We had been sent a study package months before our arrival to refresh our math skills. We reviewed the answers in the afternoon.

Friday 14 July We studied operational procedures and SOPs then had a course rules test in the morning. Flight operations were reviewed in the afternoon.

Sunday 16 July The day was spent studying the T-2 NATOPS manual to prepare for a test. I'd skimmed the manual years ago while at VT-86 but this review required more effort.

Monday 17 July The Commander, Naval Air Test Center, Rear Admiral Don Boecker welcomed us aboard – he is a great guy with vast aviation experience[158]. Skipper Rhoades gave a great welcome stressing safety and staying happy despite the workload. We had math and calculus classes in the afternoon and shared a beer at the O' Club after classes.

Wednesday 19 July I sat in T-2 and A-4 aircraft to refresh my feel around those cockpits then completed a T-2 preflight inspection for practice. We had an operations brief followed by a class softball game.

Friday 28 July Instructor, Andy and I flew an enjoyable A-4 orientation flight in the afternoon. He let me fly the aircraft after takeoff. I tried to perform a couple of 40-degree dives on a designated ground point but my roll-in was rather sloppy. I ended up too steep on my second dive. Andy demonstrated a proper 40-degree dive and my next one was much better.

Wednesday 2 August Today was our first class in Performance testing. Our pilots were sweating the load. I went over to Strike[159] and picked up a manual on F/A-18 software.

Thursday 3 August The TPS staff started explaining the mountain of work ahead of us in the upcoming year. It sounded ominous.

Friday 4 August JJ started introducing more difficult material in Pitot Statics class - the work was getting more intense.

Saturday 5 August I read F/A-18 software documentation all day long. There was a lot to learn. As the great majority of the cockpit displays were digital, there was a lot of documentation to read to understand the operations.

---

[158] Admiral Boecker is a humble hero; https://airandspace.si.edu/support/wall-of-honor/rear-admiral-donald-v-boecker
[159] Strike was the tactical jet flight test squadron at PAX. They flew A-6E, EA-6B, F-14, F/A-18 and AV-8B aircraft.

Monday 7 August We had our ejection seat training and anthropometric measurements were taken in the afternoon. Measurements of various sections of our body gave us feedback on how our bodies compare to other aircrew and aircraft cockpit design parameters.

Tuesday 8 Aug Morning classes were followed by a H-58 helicopter flight in the afternoon with Australian pilot, Miz. He is a great guy and the flight was quite enjoyable. I laughed like a kid when we lifted off the ground. We were flying but we weren't moving fast across the ground. That was a novel experience for me as I was accustomed to moving over 150KIAS to takeoff. Miz let me fly the helicopter about sixty percent of the time. He tried to teach me how to hover close to the ground with one control at a time. Flying a helicopter involved using my feet on the rudder pedals to point the nose left and right, my right hand on a control stick for movement left, right, forward and aft and my left hand on the collective to adjust power to climb and descend. I couldn't keep the helicopter stable while hovering. Coordinating my feet to keep the aircraft pointed straight ahead seemed to be my biggest challenge. Miz was a very patient instructor and returned the helicopter to a stable hover and gave the controls back to me. It was an enjoyable experience and I hoped to fly more helicopters. I got home and studied a couple of hours for a calculus test.

A VA-115 squadron mate, Bob, called from Navy Post Graduate School (NPGS) in Monterrey. An instructor-instructor crew from VA-128 Operations, Steve and Rick, were killed in an Intruder crash while practicing for an air show at NAS Whidbey Island. I was very upset to hear this. Regardless of your skill set, naval aviation can be quite dangerous.

Thursday 10 August We performed our cockpit evaluations today. It got our classmates stirred up like a bunch of lions waiting for a piece of meat. Finally, we were evaluating something. A couple of us evaluated the cockpit of the EA-6B aircraft. It was like the Intruder but had concerns that caught my attention. Those items generated paragraphs for my report. We had been taught how to create a mapping of the Field of View (FOV) outside the cockpit from our seated position. That took time to estimate the angles of the obstructions due to the canopy rails and equipment. It was exciting to evaluate another aircraft.

Friday 11 August We had a class in the morning then completed the EA-6B cockpit evaluation in the afternoon.

Saturday 12 August I went to USNTPS to pick up a T-38[160] flight manual and sit in the back seat of a T-38 to refresh my memories. My first T-38 flight was back in 1983 while at NASA. The T-38 is a great aircraft and I looked forward to flying it once again. I studied the software mechanization for the Avionics Software System Test Aircraft (ASSTA) then got to sleep at 2am.

Monday 14 August Several Systems students had a flight in the Calspan company Total Inflight Simulator (TIFS) aircraft[161] in the afternoon to evaluate the F16 air to ground radar. That was an educational flight with Duncan serving as our instructor. We flew east toward Bloodsworth Island[162] to utilize their ground-based, metal array to test the radar resolution. The array helped users quantify the resolution of surface radars in a manner like an eye test for humans. The better the resolution, the better the radar could distinguish closely spaced reflectors in the array. This translated tactically into finding small targets or discriminating between targets in a cluttered, urban environment. The F-16 radar did not match the resolution of the Intruder but it wasn't designed to do so.

Tuesday 15 August The more JJ taught Pitot Statics and I failed to study, the more confusing it became. I couldn't follow his explanations without my own study beforehand. I went home and spent all afternoon and night studying pitot statics. It finally made sense, but I didn't get to sleep until 230am. 'All of the science, I don't understand…'; Elton John came to mind.

---

[160] The T-38 aircraft has superbly served the Air Force and NASA; https://www.NAVAIR.navy.mil/product/t-38-talon
[161] The aircraft had an F-16 radar in the nose and room in the back for students to experience and discuss the technology; https://www.nationalmuseum.af.mil/Visit/Museum-Exhibits/Fact-Sheets/Display/Article/195751/convair-nc-131h-total-in-flight-simulator-tifs/
[162] https://ndw.cnic.navy.mil/Installations/NAS-Patuxent-River/About/Annexes/Bloodsworth-Island-Range/

Thursday 17 Aug We had a software class in the morning. Our Software Evaluation Test Plan (SETP) was due next Thursday. Completing it was challenging with the cockpit evaluation report also due that day. I studied for our Mechanics test, completed Pitot Statics homework and wrote cockpit evaluation paragraphs for the report. I studied the T-38 Boldface procedures for a flight the next day.

Saturday 19 August I spent the morning writing the cockpit evaluation report but it was going rather slowly. As new students trying to learn how to write reports in the standard test pilot format, we had informal access to sample reports to get a feel of how to write. Class 96 Airborne Systems student, Dee, loaned me a section of his old report. As I looked at his sample, I thought, 'This is excellent writing. I wish my writing was this good'. This thinking led to writer's cramp. I felt paralyzed to write anything. Nothing I could ever write would be as good as Dee's writing. I wasn't sure how to proceed.

Sunday 20 August The cockpit evaluation report writing proceeded very slowly. I struggled to get started but finally had a breakthrough. I resigned myself to the fact that I'd never be as good a writer as Dee, but that I needed to write from my thoughts. It didn't matter that I wasn't as good. What mattered was that I could write a report that was my own work, based on my own data and perceptions of the test. Once this mindset came, writing became more intuitive. I still needed to perfect my own style with feedback from my instructors and that came with time and practice. My writing style isn't poetic but hopefully gets the message across. As for Dee, his excellence far surpassed his gifts in writing[163].

Our systems students went over to Toad's house to work on the F/A-18 SETP. Working together was a struggle. There were too many leaders and not enough workers. I studied pitot statics, F/A-18 software and T-38 SOPs in the afternoon. I felt as though we were walking on a small path surrounded by a forest. The trees were the various equations and science being thrown at us in class. We didn't understand it well, but

---

[163] https://www.navy.mil/Leadership/Flag-Officer-Biographies/Search/Article/2236159/vice-admiral-dee-mewbourne/ and a speech at the Naval Aviation Museum https://www.youtube.com/watch?v=akJakMsjFp0

complete understanding wasn't required to be among them. The walk continued.

Monday 21 August JJ continued to painfully press on in the pitot statics class. I was not doing enough homework. We had a weapons systems brief on the TPS owned, Sergeant Fletcher tank to be evaluated. I worked on the F/A-18 SETP. Classmate and Intruder pilot, Dick and I were scheduled for a T-38 flight in the afternoon. We briefed our flight procedures then the emergency procedure of the day in accordance with the SOP. The T-38 is a smooth aircraft and feels like flying on a magic carpet when above 300KIAS.

Tuesday 22 Aug Today was our last class in pitot statics. Our test was going to be in the next week. We covered the basics of rotational kinematics in mechanics class. Performance class was running away from me. I needed to read and catch up with the work. I spent more time working on the F/A-18 SETP.

Thursday 24 Aug I turned in my cockpit evaluation report hoping for the best. Performance class was still challenging to absorb. I rewrote my portion of the F/A-18 SETP. The final product was 100 pages with photos.

Friday 25 August Communication Systems and Performance classes continued to be challenging. A T2 NATOPS quiz caught our class off guard. Quiz, no one told us about a quiz! We got ready for the USNTPS 'You'll Be Sorry' welcome aboard party thrown by the senior class for our class. The senior class explained in skits and speeches that with all the work required and pain involved, we would regret applying for TPS. While the work was very challenging and our brains hurt most of the time, we all knew that we were fortunate to be there. Despite the You'll Be Sorry title, the party itself was a great time! We looked forward to the opportunity to host the party for our junior class.

Saturday 26 August I had dinner with Beef and his family. Their youngest was a newborn and I wasn't sure how Beef was coping while attending TPS. I returned home to read the F/A-18 NATOPS manual. The work didn't stop and we needed to keep plugging along.

Monday 28 Aug Smiley provided a brief on the Missile Integration Test Plan (MITP) assignment. We had a group discussion and put together information to assist with the task.

Tuesday 29 August Performance class made a little bit more sense today. In our Communication Systems class, Dr. Masters explained that the Maxwell and Schrodinger wave equations were obvious to the most casual observer. Smash and I really loved that as those equations were not so obvious to us. I had a T-38 flight with fleet E-2C pilot, Matt[164]. After takeoff we were told that the restricted area 4005 was hot and there was a low cloud layer. We decided to practice 2.4-2.8 Gs sustained turns. That was an enjoyable flight. I got home and studied pitot statics until midnight.

Wednesday 30 August I woke up late and walked directly into the start of the pitot statics test at 8am. A couple of the problems were quite challenging. JJ was talking about Cray 5 computers in our Aerodynamics class. Andy briefed our first Qualitative Evaluation (QE) flight. Slider and I were assigned the HH-65 aircraft. We looked forward to the flight and testing. We continued working on the F/A-18 SETP.

Thursday 31 August The systems students were waiting for me to finish testing my portion of the F/A-18 SETP. I completed the testing with Rich in the cockpit of an F/A-18 in the hanger bay with power attached. The aircraft was hot and the work was frustrating. I went over to the home of one of our senior class students, Jimmy D, to discuss upcoming projects.

Friday 1 September Mechanics class was a struggle. Our T-38 NATOPS class was good and spoke about parachute gold ring failures and the requirement to pull the D ring manually if it occurred. Your heart would be pumping quickly if you were free-falling after ejecting, expecting a parachute and it wasn't opening. We had an ASTTA flight to test the F-16 air-to-air radar. We had four of us running air to air intercepts, one at a time. RIOs Beef and Weeds performed quite well and were impressive to watch.

---

[164] Matt is an exceptional officer with a quick wit and engaging personality; https://en.wikipedia.org/wiki/Matthew_L._Klunder

Saturday 2 September I completed mechanics and performance homework. Class 97 had a dinner at the Sandgate restaurant on the water. They served crab, fish and corn. A couple of the class 96 pilots joined us for a beer and discussion. Billie[165] was from Canada and Peter was from England. I enjoyed listening to their international perspectives on life.

Sunday 3 September I drove over to USNTPS to read about the HH-65 aircraft and prepare our QE-1 test plan. I worked until late. It was a beautiful day and I preferred to be outside enjoying the sunshine, but duty called.

Monday 4 September Slider and I worked on our QE-1 test plan all day. He was a very good engineer and I enjoyed working with him. I spoke with K9 and he mentioned getting out of the Navy in a year. I suggested he apply to USNTPS as the new A-12 aircraft would require test pilots[166].

Tuesday 5 September We had performance, communications, and aerodynamics classes in the morning. They all seemed to blend and create brain fog.

Wednesday 6 September We had performance, aerodynamics, communication systems classes and the SETP debrief. Slider and I finished up the QE-1 test plan. I went home to complete the F/A-18 software evaluation report.

Thursday 7 September We had communications, performance and a new class entitled 'Introduction to Controls' by Doc Richards. The last class was very interesting. Duncan briefed us on another assignment, a Weapons System Description (WSD) brief. I was given the F-14 AWG-9 radar to research and prepare a PowerPoint presentation. The oral brief was to be delivered in front of a panel of instructors while in our Dress Blue uniform. The assignment simulated the atmosphere if required to brief senior officers at the Pentagon on future weapon system

---

[165] Billie was a Royal Canadian Air Force F/A-18 pilot that later served in combat then become the lead Lockheed Martin Test Pilot for the F-35 aircraft.
https://www.youtube.com/watch?v=hWmnAxT9QcE
[166] K9 attended USNTPS, flew the Mirage in France for his DT-II final project and finished first in his class.

modifications. I continued work on the F/A-18 software evaluation report. I started work on the MITP for the AGM-65 Maverick missile[167].

Sunday 10 September I spent the day trying to catch up with all my schoolwork. There was so much to do. I started reading about the F-14 AWG-9 Radar. It was an amazing weapon system, but my lack of air-to-air radar knowledge was evident.

Monday 11 September I had my first flight in a F/A-18B Hornet. My instructor, Crutch, was a very intelligent, enthusiastic and experienced test pilot that loved the aircraft. He had been flying them since he was an Ensign[168]. Once the control tower cleared us for takeoff, Crutch added military power then full afterburners. I was impressed. The aircraft was smooth, stable and accelerated quickly. Once safely airborne, Crutch raised the gear then the flaps as we accelerated and quickly climbed away from the runway. I had never experienced a takeoff and climb-out occur that fast. We followed the departure procedures for PAX and were in the restricted area for testing within five minutes.

Crutch spoke quickly as he tried to pack as much information into our flight as possible. He demonstrated the air-to-air radar and how an intercept was conducted using other aircraft in the restricted area. He enjoyed showing me the great turning capability of the Hornet as he snapped on a 7.3G turn. It was my first acceleration over 6.5Gs and I could feel the difference. It was a great turning aircraft and didn't bleed all its energy as an Intruder did. The Hornet's two afterburning engines helped it regain airspeed quickly when needed.

Crutch added power to take the aircraft into supersonic flight. I'd never been supersonic and was surprised that the aircraft handling didn't change as we passed through Mach 1 and into M1.1. The aircraft was very smooth and had very impressive handling qualities. We returned for landing as Crutch lowered the flaps and landing gear. The aircraft appeared very easy to fly compared to the Intruder. The flight opened my

---

[167] The Maverick missile; https://man.fas.org/dod-101/sys/smart/agm-65.htm
[168] Crutch had enlisted in the Navy but was quickly picked up for an officer program. During his officer candidate training, he was accepted to attend flight school where he excelled once again.

eyes for things to come. It was a great blessing to have Crutch as an instructor.

Wednesday 13 September We delivered the F/A-18 Software brief to our instructors. The instructors complimented us as being very professional in delivering the brief.

Friday 15 September Classmate Marine Harrier pilot, Rider and I flew a cross country in a TA-4J to Pensacola, FL to record instrument time. Rider was very calm, kind and funny character and a pleasure to fly with. We took off from PAX just before sunset. As we climbed up to our cruising altitude above 30 thousand feet, there was a row of thunderstorms along the east coast of the USA. They were lighting up the sky and making impressive scenery for our trip south. Unknown to us at the time, Fayetteville, North Carolina was getting smashed by these storms[169]. Once Rider and I were at our cruising altitude and following our navigation aids, we had time to watch the storms that extended for hundreds of miles along the coast. Lightning bolts were occurring every few seconds. We flew for 2.4 hours and landed after 9pm at NAS Pensacola.

Tuesday 26 September I arrived at USNTPS imagining that my writing skills were well developed through assignments at university and in naval fleet settings. My Cockpit Evaluation report was returned full of red corrections as if someone had bled all over it. I had a good debrief with Crutch and was told this was normal for a student's first effort. As time passed, my writing improved.

Thursday 28 September Classmate, Spot and I had a T-2C spin flight. He was a fleet A-7E Corsair pilot who was engaging, enthusiastic and articulate. I hadn't been in a spin since my first flight in the T-34C. The T-2C spin was anti-climactic as I anticipated more dramatic maneuvers. Spot was a great pilot and it was a pleasure to share the cockpit.

---

[169] This report covers the storm; https://pubs.usgs.gov/wri/1992/4097/report.pdf.

Monday 2 October I gave the WSD brief for the F-14 AWG-9 system. The instructors were satisfied with the presentation and fortunately didn't ask too many difficult questions.

Tuesday 3 October The helicopter pilot, Slider and I flew the HH-65[170] Dolphin radar evaluation flight. The radar worked as designed but the performance was disappointing. If you were stranded at sea in a survival situation, you would want an aircraft with a better radar.

Wednesday 4 October I had a T-2C turn performance flight with Randy. He was a soft-spoken yet perpetually helpful pilot with a fleet P-3 background. Later that night, Beef, Weeds and I had a UH-60[171] NVG flight piloted by Kidd. This was our first flight using NVGs. We were truly amazed as they turned night into day. We hovered near a grove of trees but a few feet off the ground. We rose above the trees to simulate firing a weapon then reset behind them for protection. We instantly saw the tactical advantage of NVGs in low altitude operations.

Thursday 5 October A F-16XL design discussion with Powell in Performance class was interesting. An F-14 study was mentioned that suggested the aircraft be used as an attack platform[172].

Friday 15 December Crutch and I tested the F/A-18 radar at the Bloodsworth Island radar array. The Hornet needed to offset the target from the nose of the aircraft due to the pulse-doppler ground mapping radar. The resolution was good but didn't match the Intruder's radar.

---

[170] The HH-65 helicopter; https://www.naval-technology.com/projects/hh65-dolphin/
[171] The UH-60; https://asc.army.mil/web/portfolio-item/black-hawk-uhhh-60/
[172] The F-14 did become the Bombcat and served in Bosnia and Afghanistan during Operation Enduring Freedom after 11 September. https://theaviationist.com/2020/12/26/how-the-f-14-tomcat-evolved-and-became-the-bombcat/

# Chapter 10

USNTPS Senior Class 1990 and Tornado Testing in Italy

*There was rapidly rising terrain to 4100 feet MSL in a 90° left turn with a dark radar screen at 1nmi. This indicated that we had nine seconds until terrain impact.*

Thursday 11 January TPS instructor pilot, TP and I completed an inverted spin demo in a T-2C. I'd done upright spins in my first flight in a T-34 and in a T-2C with Spot but had never been in an inverted spin. TP was a calm, easy-going and reassuring fleet Intruder pilot that made everything appear simple. The brief covered the maneuvers and the recovery procedures. The flight was simple enough minus what the aircraft might do as we flipped around the sky. We tightened our lap belts for the inverted spins as they generate negative Gs and we didn't want to smash our helmets on the canopy. I was apprehensive going into the spins, but they were more benign than anticipated. We finished doing a few inverted spins and landed back at PAX leaving me with the thought, 'No big deal'.[173]

Friday 19 January Res and I shared our first flight together in a T-2C. It felt unusual flying with him after spending a great deal of social time together. I wondered if we would struggle to communicate airborne despite his strong English language skills. Aviation has its own language and we communicated quite well. Res was an excellent pilot, and I was privileged to fly with him. Res also had a newborn daughter to add challenge to his life.

Monday 12 Feb The Systems students were flown down to Cherry Point, NC to perform a QE on the A-6 radar and FLIR using a TC-4C trainer aircraft. As an Intruder BN, I helped the other students understand the switchology for the computer, INS, search radar and FLIR. The other students weren't impressed with the multitude of switches, knobs and buttons and wrote deficiency paragraphs citing the burden they generated for the BN. They were correct but BNs were well trained to compensate for the complexity of the system. It was a gratifying experience to share

---

[173] This assessment of inverted spins changed after a June 1990 flight in a Pitts, open cockpit biplane.

the Intruder cockpit design with crews with other aircraft backgrounds and listen to their comments.

Friday 16 February Class 97 finally earned the opportunity to host Class 98 for their 'You'll Be Sorry' party. As we drove up to the O' Club, a large surprise appeared on the front lawn. Classmate Dick had borrowed a 35-foot tall, inflatable King Kong gorilla from a friend and placed a sign on its belly; 'You'll Be SORRY!! Class 98'. That was very funny and set the tone for the party! Our class skits came off well and the junior class enjoyed their last festive moments before the real work began. We were happy to have that social hurdle behind us with a busy schedule ahead.

Monday 12 Mar As spring arrived, flights in our glider aircraft raised the class excitement levels. Many of us had never flown in a glider and were looking forward to the opportunity. We towed the X-26[174] gliders with a U-6A Beaver[175] aircraft over to Navy Webster Field, 10 miles south of PAX. Webster was used for various testing but didn't have the flight restrictions of a control tower due to light airborne traffic. I had a short flight with Marine Instructor, Andy and a longer one with classmate, Spot.

To prepare for the launch of the X-26, we positioned it on the runway by lifting each wingtip and rolling it into position. We attached the tow rope loop into the glider hook on the forward bottom of the aircraft. We got into the aircraft and strapped ourselves into the harness. The Beaver was directed by ground personnel to take the slack out of the rope and we were ready for takeoff. The Beaver pilot confirmed our readiness for takeoff over the radio, added power and we started rolling. We built enough speed to fly in 15 seconds, the pilot gently pulled back on the stick to lift off. We stayed low in ground effect waiting for the Beaver to achieve takeoff speed then both climbed to three thousand feet MSL.

Once we achieved our desired altitude, the pilot released the rope with a control in the front cockpit and we were free of the Beaver. The Beaver returned to land in preparation for the next aircraft tow. Flying the

---

[174] The X-26A glider; https://www.NAVAIR.navy.mil/product/x-26a-frigate-glider
[175] The U-6A; https://www.NAVAIR.navy.mil/product/nu-1b-otter-u-6a-beaver

glider was a great experience. Without an engine, the X-26 was very quiet. The only noise was the wind rushing over the canopy and our crew discussions over the headset. Even the airflow noise stopped when we were upside down at the top of a loop and when the airflow over the aircraft was minimal. The wind came rushing back as we accelerated back down to the bottom of the loop. Our typical flight time in the glider was about 20 minutes as there weren't thermal updrafts to keep us airborne. We spent several hours at Webster Field before heading back to PAX.

Saturday 17 March Class 97 traveled to Edwards Air Force Base for a field trip to explore how the Air Force conducts flight test on the west coast. Air Force TPS sent their students to PAX earlier in the year for a similar exchange.

Sunday 18 March Class 97 went out to Tehachapi airport to fly gliders followed by a Mexican dinner on the way back to Edwards AFB. I was fortunate to fly in a two-seat glider, the Blanik L-13[176]. Tehachapi sits amongst the Sierra Nevada mountains and enjoys thermal updrafts from the surrounding topography that make for good gliding. We received a tow from another aircraft up to 3000 feet AGL then started chasing the thermals. At times, we were able to get 500 feet per minute in a climb due to the updrafts. As it often turns out, the thermals were in a rather narrow cone of sky above a fixed point on the ground. This led to several aircraft sharing the cone as we all tried to gain altitude and extend our flight time. The proximity of the aircraft resembled a graceful ballet.

Unfortunately, the updrafts dissipated, and we descended without recourse. My pilot set up a good position abeam his landing point with enough airspeed and altitude. That is a key for a successful landing in a glider. He landed beautifully on the runway.

Tuesday 20 Mar While on our field trip to Edwards AFB, Crutch coordinated a night, F/A-18D NVG flight from NAS China Lake[177] for me. He knew that NVGs were going to be used extensively in tactical aviation in the future and especially in the upcoming A-12 aircraft.

---

[176] The Blanik L-13; https://en.wikipedia.org/wiki/LET_L-13_Blan%C3%ADk
[177] NAS China Lake is a one-hour drive north of Edwards AFB.

Crutch knew an experienced NVG pilot[178] at China Lake that was happy to provide a demo. We briefed the flight with the explanation that if we needed to eject, the call would be, "Goggles, goggles, goggles, eject, eject, eject!" This was necessary as the engineers predicted that aircrew would break their neck if they ejected wearing NVGs. They were relatively heavy, mounted in front of the helmet and the acceleration of ejection would place a fatal load on an aircrew's neck. As the moon wasn't above the horizon for our flight, we used illumination from stars to light our path.

We took off normally then put on the goggles when clear of the airport to the east. Wearing the goggles turned the night into day. The NVG green cockpit lighting in the F/A-18D was compatible with the goggles allowing us to easily see out of the aircraft or scan inside for flight data. Many other fleet aircraft used red lighting for night flight. These lights resulted in large light blossoms while viewing them with NVGs making them unsuitable for use. We flew east to the Panamint Valley, headed north while descending to 200 feet AGL. An altitude this low was usually only flown during VFR, daytime conditions. We maintained this height above the ground throughout the night low level despite rugged terrain, narrow canyons and busy cockpit taskings.

As we flew lower, the resolution of the goggles improved. This was a risk for aircrews as they were tempted to descend below 100 feet to improve the resolution. Depth perception on the goggles wasn't as good as the normal eye in daytime so there was a risk of collision with the ground. The radar altimeter was set for 190 feet and helped keep us honest and safe. We flew along the flat terrain at 1500 feet MSL in the valley for 5 minutes, climbed over a six-thousand-foot ridgeline then descended back to 1500 feet MSL into the Saline valley.

At the northern end of the Saline, we turned to the west and cleared an eight-thousand-foot MSL ridgeline. We descended back down to forty-two hundred feet MSL with a turn to the south into the Owens Valley. After passing the small town of Lone Pine, we added power, commenced a turn to the west and started a climb to ten-thousand feet MSL into the Sierra Nevada mountains. After a few minutes flying along

---

[178] Unfortunately, my logbook doesn't list the pilot's name for this flight.

a high-altitude plateau, we turned south into the Kern River Valley. It was here that my mind was forever opened to the power and flexibility of NVGs. We were tucked down into a rather narrow valley with ridges above us on both sides. While at just a few hundred feet above the ground at 400KIAS, the pilot completed several simulated long-range, air-to-air missile shots at the airliners outside our restricted area.

This was a revolutionary concept. Flying this low at night, in such a narrow valley with rugged terrain was not safely possible with the SRTC of the Intruder. With NVGs, the pilot was flying more tactically while prosecuting air targets simultaneously. I was amazed and expressed that to the pilot. We returned to China Lake with a great appreciation for the flight and the pilot's skill! Crutch was present for the debrief to see my reaction to the flight. I thanked him very much for the opportunity to experience this shift for tactical aviation.

Monday 2 April The Systems students had a field trip to the Baltimore airport to fly aboard a BAC 1-11[179] ground mapping radar technology demonstration aircraft. The aircraft used Synthetic Aperture Array (SAR) technology[180] to produce very high-resolution images. We were all very impressed. This type of radar technology was planned for many of the newest aircraft including the A-12 under development.

Tuesday 3 April Class leader, Mike and I were getting ready for our DT-II assignments and shared a TA-4J flight to practice our test techniques. He was a fleet P-3 pilot so flying this sports car of an aircraft was a great experience for both of us. As we had both flown the A-4 quite a bit, we knew the results of the testing. In the testing area, we completed G awareness maneuvers then wind-up turns. These turns increased the G loading while maintaining airspeed resulting in a tighter and tighter nose down spiral until we stalled the wing or reached our maximum G limit[181]. The turns provide various information about the aircraft including stick force per G and accelerated stall behavior. The Skyhawk had an incredible roll rate of 720° per second but we were limited by NATOPS

---

[179] Westinghouse Bac1-11 testbed; https://photorecon.net/northrop-grummans-final-mission-in-the-bac-1-11/
[180] SAR technology; https://en.wikipedia.org/wiki/Synthetic-aperture_radar
[181] The USNTPS Flight Test Manual;
http://www.usntpsalumni.com/Resources/Documents/USNTPS_FTM_103.pdf

to a 360° roll to avoid flight control issues! We set ourselves up for a 360° roll; three, two, one, mark and Mike snapped on the roll. Yes, it was very fast (less than a second). We had to settle our brains after each roll.

We finished more testing and had enough fuel to simply enjoy the aircraft before landing. As luck would have it, there were great cumulus nimbus clouds in the area that formed various caverns for exploration. Mike turned the aircraft over to me to fly from the backseat. We raced around the clouds weaving in and out of holes, canyons and mountains. At one point Mike made the comment, "Use the force, Luke"[182]. We were having the time of our lives chasing around the skies. The flight brought us great joy while preparing us for the testing ahead.

Tuesday 10 April I had my first flight in a F-14 Tomcat. The flight was a Qualitative Evaluation 2 (QE2) with fleet Tomcat pilot O'Brien. He had flown the bird up from NAS Oceana and parked it on the TPS flight-line. I was to evaluate the aircraft for its suitability to perform the Strike Fighter mission. At this time, the Tomcat hadn't been dropping bombs in the fleet as its clearance was limited due to past testing constraints.

In preparation for the flight, I studied the aircraft's systems including the AWG-9 radar. I briefed O'Brien on the order of the tests. I planned to conduct maneuverability tests to assess the aircraft's agility then test the radar for detection and tracking functions. O'Brien briefed me on the unique crew coordination items for the Tomcat, we visited maintenance control then walked out to the jet.

The first thing I noticed about the Tomcat was its large size. It was massive for a carrier-based, tactical aircraft. Once in the cockpit, I was impressed by the large amount of open space compared to the TA-4J or F/A-18. It felt like a big living room. The weapon system avionics interfaces were aged and worn but that didn't bother me as an Intruder BN. I flipped the switches required to get the avionics warmed up properly. O'Brien started the engines. They had a deeper, more powerful

---

[182] In November 2009, I turned on the TV to see Mike completing a space walk outside the Space Shuttle while supporting the International Space Station. He was selected to be a Mission Specialist astronaut in 1998 and was certainly 'using the Force'! https://www.nasa.gov/sites/default/files/atoms/files/foreman.pdf

growl than Hornet engines. As we started to taxi, I immediately sensed the weight of the beast. It was like driving a Cadillac limousine. We made our way to the runway and were cleared for takeoff[183].

O'Brien selected Zone Five afterburner and we accelerated down the runway. It didn't accelerate as fast as a Hornet. It was nevertheless a thrill to be rolling down the runway very smoothly. Once airborne, that smooth feeling continued as we accelerated away from the ground, the gear came up and locked, flaps came up and the wings swept back. I was impressed and wrote down data from the takeoff. We proceeded out to the testing area and set up for the agility tests.

I'd been spoiled by doing 360° rolls at 720° per second with the A-4, but the full stick deflection roll rate of the Tomcat with wings swept didn't even match what I recalled from the Intruder. That was a surprise. We conducted the AWG-9 radar tests and the system excelled with detecting and tracking targets at long and short ranges despite aggressive maneuvering. I was impressed with the range of the radar and wished the Hornet had that capability. We returned for landing and once again the Tomcat was graceful on approach and touchdown. We taxied back to TPS and debriefed the flight. I thanked O'Brien for his assistance and began writing paragraphs for the QE2 Report.

Friday 13 Apr An instructor came into our class to offer a special deal. The two-seat P-51 Mustang aircraft, 'Crazy Horse[184]', that the fixed wing pilots were using for performance evaluations was being flown back to Orlando, Florida the next day with an open back seat. Operations wanted to know if there was anyone that wanted to ride in that back seat to Orlando. If so, they would be responsible to pay for their own commercial transportation back to PAX. I won a coin toss, was selected for the flight and started coordinating transportation back to Washington, D.C. from Orlando. I had an afternoon flight in a T-2C with 1Mig and went home to pack an overnight bag.

---

[183] This aircraft was a F-14A with the older TF30 engines. The F-14B/D had General Electric (GE) 110-400 engines that provided more thrust and performance.
[184] Crazy Horse is available for commercial joy flights; https://www.stallion51.com/mustang-flight-ops/crazy-horse/

Saturday 14 Apr I met Lee Lauderback[185], the pilot and owner of Crazy Horse, in the morning at Base Operations. We filed a two-leg flight plan to get to Orlando as we needed to stop for fuel along the way. We made our way over to the flight ramp at TPS. Being a Saturday, we were the only ones there. Lee had me strap in the back seat as he completed the preflight inspection. He started up the beautiful motor with its wonderful, deep idle and we taxied shortly afterwards.

Once we received our takeoff clearance from tower, Lee added power smoothly to full and we were airborne quickly. He kept the aircraft low over the runway until near the departure end then pitched the nose 30° up as we departed in a colorful manner. Once stable in the climb, Lee turned the controls over to me. The aircraft was wonderful to fly. The longitudinal trim control wheel had a very tight steel cable wound around it and provided very precise pitch control. As we prepared to land the aircraft at Myrtle Beach, NC for fuel, I happily turned the controls back over to Lee. Lee expertly landed the aircraft and we taxied over to the ramp for fuel.

Once refueling was complete, we picked up the second leg of our flight plan and requested clearance for takeoff. Lee took off again with a more standard climb away from the ground then returned the controls to me. We flew down to Kissimmee airport near Orlando and landed. Lee parked Crazy Horse next to Arnold Palmer's private jet in his hangar. I flew to D.C. from Orlando airport in the morning, rented a car and dropped it near PAX. I felt blessed to experience Lee's flying skills and Crazy Horse!

Monday 16 Apr Rider and I had an enjoyable flight in the X-26 glider. Smiley and I flew from Webster to PAX in the U-6A. It was funny to be flying with him in such a slow-moving aircraft after all our time crewed together in Intruders. Naturally, his flying skills were superb and he loved flying the aircraft. The Beaver is a classic piece of aviation history.

---

[185] Lee is an exceptional pilot with a long history of service including being the Chief Pilot for golfer, Arnold Palmer; https://www.stallion51.com/mustang-flight-ops/lee-lauderback/

Tuesday 24 Apr Today's flight was unlike any other. It was in a specially modified Learjet that allowed the engineers to change the flying behavior of the aircraft while airborne enabling the demonstration of variable stability. A pilot always likes an aircraft to do what he or she commands with the flight controls and not otherwise. That is a simple concept. However, dynamic systems like aircraft can exhibit motion that is oscillatory and not smooth (well damped). The oscillatory motion can be abrupt and interfere with a pilot's control inputs. Aircraft motion damping can be too restrictive or not restrictive enough. This can all lead to aircraft that are very difficult, if not possible, to control. The Variable Stability Learjet[186] allowed us to experience the controls of an aircraft with poorly designed stability.

While in the copilot's seat, I flew the Learjet with its well-designed, standard flight controls. It was a sweet ride with controls that had a very nice feel. Once accustomed to the controls, the training began. The onboard engineer digitally changed the response and damping coefficients making the aircraft quite difficult to fly. I started laughing loudly at how badly the beautiful jet could fly with the wrong flight control design. The flight was an eye opener for aircraft design but the lesson applied to all dynamic systems.

Monday 30 April Our final evaluation assignments (DT-II) were given to the class. The task was to complete a comprehensive test of an aircraft we had never experienced in a series of four flights. Everyone was nervous to see what aircraft they would be assigned. I was given the Tornado GR MK1 aircraft to evaluate for the all-weather attack mission. The aircraft was to be flown from Pratica Di Mare Airfield, south of Rome, Italy. I knew very little about the Tornado other than what I'd read years before in military aviation books. It was a tandem seated, swing-wing, supersonic strike aircraft flown by Italian, German and British pilot/bombardier crews.

Once given our assigned aircraft, the clock started for the research and test plan writing process. We had a week to write the test plan. An aircraft description that included the weapon systems was due as part of the final report two weeks after all flying was completed. My testing

---

[186] Calspan teaches pilots in their modified Learjets;
https://calspan.com/aerospace/advanced-flight-test-training

required an in-depth understanding of the aircraft. I went to the TPS library and found an unclassified flight manual for the Tornado. It was another thick flight manual to digest along with memorizing operating and boldface emergency procedures, prohibited maneuvers, airspeeds and engine limitations.

Tornado aircraft like the one flown for my DT-II final exercise[187].

Thursday 3 May Classmate Pappy and I had a T-38 flight to practice DT-II data collection.

Monday 7 May Submitted my DT-II test plan for the Tornado. I hoped it was acceptable as there wasn't much time to change it. We were leaving for Rome in 2 days.

Tuesday 8 May Smash and I had a T2C flight to practice DT-II data collection.

---

[187] Photo credit Aldo Bidini;
https://commons.wikimedia.org/wiki/File:Panavia_Tornado_IDS,_Italy_-_Air_Force_JP6706853.jpg

Wednesday 9 May Crutch and I left TPS at noon for Dulles airport, outside of Washington D.C., to board a flight to Rome, Italy.

Thursday 10 May Crutch and I arrived at Rome International airport at noon. It is located fifteen miles to the southwest of central Rome. We drove via a rental car to our accommodation located ten miles to the southeast of the airport. Our modest hotel was across the street from the ocean. I spent most of the next ten days studying, reading, writing and sleeping rather than enjoying the scenery.

Friday 11 May Crutch and I drove fifteen minutes to the military airbase, Pratica Di Mare. We were hosted by the Italian Flight Test Wing (Reparto Sperimentale di Volo – RSV)[188]. The Test Wing was the Italian center of excellence responsible for experimental and operational flight test for aircraft, software, hardware and weapons. Their flight crew and engineering staff included graduates from several test pilot schools. Italian astronauts have previously worked at the Wing. Everyone was exceptionally friendly and we quickly felt comfortable.

We were given a local course-rules briefing. The single runway was 9800 feet long x 151 feet wide in the 13/31 direction paralleling the Mediterranean coast a mile away. The traffic pattern was at 1000 feet and over the water. The military restricted area, IR62, was about 40nmi to the southeast, taking less than 10 minutes to commence our performance testing. I had three low levels planned through the Italian countryside: short, medium and long in duration. These would permit evaluation of the low altitude flying qualities of the aircraft along with the terrain following radar.

I was provided a Tornado in the hangar to complete three hours of ground testing. This allowed my flight manual study to be put into actual hardware testing. The temperature was a cool 50°F with a light breeze blowing. I aligned the IMU, selected Operate and collected INS drift data for a couple of hours. Ideally, the IMU shouldn't drift at all and maintain

---

[188] A website for the Test Wing can be found here. https://www-aeronautica-difesa-it.translate.goog/2023/05/04/reparto-sperimentale-di-volo-missione-addestrativa-conclusiva-per-i-nuovi-display-pilot-e-display-crew/?_x_tr_sl=it&_x_tr_tl=en&_x_tr_hl=en&_x_tr_pto=sc

the position as the aircraft was stationary. Unfortunately, friction caused the IMU to drift more than I expected. This played out on the first flight.

Monday 14 May Crutch and I were flown in a large transport aircraft southeast to the Gioia Del Colle airbase[189], the homebase for the operational Tornado fleet squadrons. We shared a training session in their simulator. While at the base, we were fortunate to witness an airshow by the Italian flight demonstration team, Frecce Tricolori[190]. They flew the Aermacchi MB-339 aircraft and were amazing to watch. They released tri-colored smoke during their show representing the Italian flag. We flew back to Pratica Di Mare in the transport aircraft relishing an amazing day!

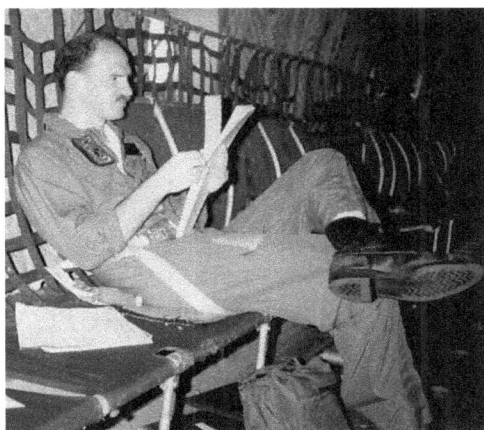

Studying Tornado procedures for the simulator. Photo by Jeff Crutchfield.

Tuesday 15 May Crutch flew the Tornado in the morning. While he was flying, I briefed my Italian test pilot, Alberto, for our afternoon flight. Alberto was fluent in English making the job easier. We planned to fly out to IR62 and collect performance and maneuvering data. I had kneeboard cards and a portable cassette voice recorder to collect the data[191]. Once the initial data was collected, we planned to complete the short low-level flight at 420 KGS.

---

[189] Gioia Del Colle; https://skybrary.aero/airports/libv
[190] Italian Flight Demonstration Team; https://www.wantedinrome.com/news/frecce-tricolori-a-short-history-of-italys-aerobatic-jets.html
[191] Despite the years, the cassettes still work and provided the following details.

It was emphasized at TPS that we should take advantage of every moment airborne to collect test data. If we had only one flight in our designated aircraft due to maintenance or other issues, the final report would be written on that single flight. With that in mind, I was anxious to test the radar. I was confident the Tornado aircraft performance would impress me, but I wasn't sure about the radar. The radar would be fundamental in finding discrete targets and flying at low altitude in all weather conditions. If it wasn't well designed, the aircraft wouldn't be well suited for the all-weather attack mission.

While I was working on the radar, I gave Alberto various flight profiles to collect data on the fuel flow, range and endurance of the aircraft. He handled all the communication with the air traffic controllers for our flight. Most of the communication was in English and understandable. At times the controllers spoke very quickly in Italian. I relied on Alberto to translate and notify me if it impacted on our mission.

We preflighted the jet then strapped into our ejection seats. I started the IMU alignment and Alberto got the engines started. The INS alignment finished and we called for taxi. Just before takeoff, we recorded the fuel load as I turned on my cockpit audio cassette recorder. I started my stopwatch as Alberto added military power. As we didn't use the Combat afterburner power setting, it took close to a minute to lift off. We wanted to save our fuel for other testing on this flight. Alberto climbed up to 500 feet MSL for our transit to IR62 to the southeast. As we took off, I immediately noticed the smooth, stable feel of the Tornado. It improved as the speed increased and the wings were swept back generating a large wing loading. It was that smooth 'magic-carpet' feeling, like the T-38 from 300-400KIAS. Alberto set our maximum range airspeed and reported our fuel load and flow rate.

I started the radar assessment shortly after takeoff. The resolution appeared adequate for fine targeting in close but the long-range performance was less than expected. On arrival in IR62, Alberto asked ATC for a block of airspace from 1500-20000 feet MSL. Our request was approved and we executed a maximum range climb from 5000-20000 feet MSL. Alberto called out the indicated airspeed, fuel load, fuel flow and angle of attack every 2000 feet while I wrote down the data along with an elapsed time. Upon reaching 20000 feet, a final fuel load was recorded and Alberto set the power for cruise. As we reached the boundary of IR62, we recorded our fuel load, airspeed and reversed course.

We conducted aileron rolls to assess the performance and stability of the doppler and search radars with a 45° wing sweep. The radar suffered performance degradations then had a failure that couldn't be corrected airborne. Further radar testing was not possible. We shifted the testing to aircraft maneuvering performance. Alberto conducted instantaneous turn performance tests at 15,000 feet MSL with 25° and 45° wing sweeps in military power while I recorded the data. We lowered to 5000 feet MSL and repeated the 45° wing sweep turn. The data provided a glimpse of the maneuverability of the aircraft in simulated combat conditions. We continued our descent down to 1000 feet AGL and entered our short low-level route to assess the aircraft's response to turbulence and wind gusts at 420 KIAS with a 45° wing sweep.

The Tornado was a smooth ride. Combat aircrew would love the comfortable flying qualities of the aircraft as it allowed them to focus more on their mission success along with defeating enemy defenses. Alberto set up an acceleration demonstration from 250-600KIAS using Combat (full afterburner) thrust at 1000 feet using variable wing sweep. We recorded the fuel before and after the acceleration run. It took a bit of fuel, but the acceleration was impressive. Attack crews would love that performance!

Alberto checked in with Pratica Tower and we came in for a left overhead break to the downwind. We completed a touch and go landing and returned for a full stop. The aircraft was smooth in the landing pattern. Post flight analysis showed poor performance of the IMU. It was replaced with another unit prior to our next flight. We debriefed and looked forward to better radar results on subsequent flights.

Crutch and I discussed our flights over dinner. He was flying back to the USA and leaving me to complete the DTII flights. I'd appreciated all his help in setting me up so far from home. I headed back to my room and continued writing paragraphs for the final report.

Wednesday 16 May My longest flight in the Tornado (1.4 hours) occurred today. While Alberto spoke in English to the controllers on the first flight, it was mostly in Italian for the remainder of the flights. That was a good thing as he and the controllers could communicate quickly and more succinctly in Italian. Alberto kept me updated on the flight path approvals and restrictions. The preflight, startup, alignment and taxi all occurred without incident. The takeoff utilized full afterburner thrust. The aircraft accelerated quickly down the runway and we were airborne in no time. We continued our acceleration to 250KIAS then Alberto reduced power and initiated a climb to 1000 feet MSL at 270 KIAS for our transit to IR-62.

I resumed my assessment of the ground mapping radar resolution. I was able to use cultural landmarks and farming irrigation ditches in my examination. The radar provided caution lights indicating it was having problems, but it continued working. I looked at ships at sea then we conducted a flyover navigation update to examine IMU drift.

Once inside IR-62, we were initially limited to 12,000 feet MSL due to other traffic. We climbed up to 12,000 while I continued to examine the radar performance. We collected various performance data then conducted a combat thrust climb from 3000-20000 feet MSL with 45° wing-sweep. As the aircraft was light and without external stores, the climb took less than a minute. There was no doubt that the Tornado was an impressive aircraft.

We gradually descended back down toward the water in hopes of testing the Terrain Following (TF) system. Alberto suggested we fly manual TF just inland of the beach as the mountains on our proposed route north of Gaeta were getting clobbered with thunderstorms. I agreed and we set up the system for TF. The air traffic controllers had us hold our position for a few minutes then the TF system wasn't performing well. When a TF failure occurred, the system had a loud 'WOOP' audible alarm with red flashing lights. It got my attention as it was very loud and annoying in my opinion. Aircrew appreciate the fact that they are low and fast and that failures can be lethal. However, the lights and alarm were excessively intrusive. We obtained overland TF data while flying at 600 feet AGL. However, the system was not performing correctly and Alberto was disappointed.

We steered toward Ponza Island to evaluate the TF system over the water. We could select Soft, Medium or Hard Ride on the automated terrain following system and the aircraft would adjust the flight profile accordingly. We set up TF automatic, soft ride at 300 feet AGL and 400KIAS. As we came toward Zannone Island enroute to Ponza, the system commanded a smooth pullup to clear the island crest that rose to 636 feet MSL. Once we cleared the terrain, the system commanded our descent. That worked well and was impressive to experience. We headed back to Zannone Island for another TF experiment but in Hard Ride mode.

We stepped down to 200 feet AGL and maintained 400KIAS. As we approached Zannone Island, the system commanded a brief 3G pullup to clear the terrain. Once clear of the crest, the system commanded our descent back to 200 feet. That was impressive! The TF system in automatic mode performed well and the aircraft was comfortable at low altitude and high speed. We concluded our testing and returned to Pratica Di Mare for our debrief. Alberto hoped to demonstrate the automatic TF system in mountainous terrain on the next flight.

Thursday 17 May Alberto and I briefed the flight starting with a radar assessment pass at Pratica followed by a low level at 450KTAS through the mountains and valleys to the northeast concluding with a simulated attack on the Grazzanise military airbase. We planned to finish the flight with practice level laydown and loft weapon delivery profiles to assess the aircraft handling and weapon system software.

Alberto used minimum afterburner for takeoff. We accelerated to 250 KIAS once airborne, performed autopilot disconnect checks and turned back into the field at 9nmi and 1000 feet MSL. I examined the radar resolution using runways, taxiways, hangars and aircraft.

Once overhead Pratica, Alberto turned right to the city of Valletri, swept the wings to 45°, added power to set 420KTAS while descending to 1000 feet AGL. We passed Valletri then Artena, increased our speed to 450KGS and started using manual TF with heading command and a 500-foot clearance setting. The TF system worked well. We approached the mountains near La Forma and entered the Parco Naturale Regionale Monti Simbruini[192].

Alberto engaged automatic TF and set 500 feet AGL, Hard Ride for our terrain clearance. The TF at times calculated that our clearance was going to be below the set threshold. The two big red lights started flashing on either side of the radar screen and the loud, 'WOOP' sound blasted in my headset. It was startling and excessively loud. We were not about to hit the ground and die, we were just a little low. The aircraft was overreacting in a way that added stress unnecessarily. This warning didn't happen just once but repeatedly as we slipped slightly low. It was quite distracting and could prevent aircrew from concentrating on bigger threats.

We passed east of the mountain town of Vallepietra enroute toward Capistrello. As we approached Monte Tarino at 6450 feet tall, clouds obscured the peaks. Alberto disconnected TF, climbed to ensure we stayed VMC with an escape heading to lower terrain if weather mandated it. The clouds provided a safe exit if needed and we remained on course.

A few miles short of Capistrello, Alberto made a hard 90° turn to the right and descended into a valley leading to a bridge near Sora. Once out of the turn, he reengaged auto TF. As we raced along the valley at 500 feet AGL, 2000 feet MSL, the mountains soared to 6700 feet MSL just to our right. Cars were travelling on the expressway alongside us. I suspect a few would have caught a glimpse of us racing past. Adjacent to Sora, we steered slightly left toward Monte Lungo. We sidestepped the town to the south and proceeded next to the ridge. The TF continued to perform well. We headed toward Grazzanise air base but they had aircraft in their traffic pattern so we were unable to proceed on our simulated attack. We diverted over the Gulf of Gaeta to conduct our practice attacks using a tower near Sperlonga as the target.

We set our nose on Ponza Island to gain distance from Sperlonga and climbed up to 3000 feet MSL. We turned into Sperlonga at 17nmi

---

[192] The scenery was stunning as the mountains rose to 6700 feet tall; https://translate.google.com/translate?hl=en&sl=it&u=http://www.parcomontisimbruini. it/&prev=search&pto=aue

while descending to 1500 feet AGL. We completed the practice attack and headed back toward Ponza for another run. We turned back into the beach again at 14 nmi for a loft attack and engaged TF once inbound. The mission computer commanded the pullup at the appropriate range and the system completed the simulated release.

We turned back outbound from the beach to set up another pass. We turned back toward the beach for a high loft delivery using a 3G pullup. This maneuver was like throwing the bomb into the distance and enabled crews to remain away from a target with anti-aircraft threats. We finished our analysis of the attack software and returned to Pratica for our landing.

Friday 18 May My fourth and last flight was to occur today. It was my fourth flight but a bird made it not my last. Alberto and I briefed the flight that would include the medium length low level to the northeast of Rome. Alberto set the power at military for takeoff then increased the power to Combat thrust at 150KIAS. He deselected Combat power at 250KIAS as I provided steering to our first navigation point, 31 nmi to the northeast.

A few moments later, a bird flew toward the port side of our aircraft. We had been airborne less than two minutes. Alberto checked the engine instruments, reported that they appeared normal and suggested the bird missed us. We were set to proceed on the flight when he noticed that the port engine was running 100°C hotter than the starboard engine with the identical power setting. That was abnormal and we returned for landing.

Alberto commented in the debrief that it had been the first bird-strike of the season for him. We had been airborne less than 10 minutes. Alberto felt terrible that my Tornado flight time had been compromised by various problems. In the debrief, he noted that I'd not sufficiently experienced the Tornado fully automatic TF mode over land and that we would reschedule the long low-level route for Monday. I was in no position to argue as a guest and looked forward to a 5th flight.

Monday 21 May Alberto and I briefed our low level to the northwest of Rome. The route had eight turn-points with 2-3 minutes between each. It was another beautiful day at 72°F with light winds from the south. After takeoff, I completed more examination of the radar while Alberto coordinated with tower for our transit north to Passo Carese.

Once outbound from Pratica, Alberto coordinated in Italian with Champino Approach Control to pass overhead the old Rome international airport, located eight miles to the southeast of central Rome. This was a priceless opportunity to experience flying in a tactical military jet across the Italian countryside. We flew for a couple of minutes past Champino to pass clear of the urban sprawl of Rome before descending for the low level. Alberto swept the wings to 45°, engaged automatic TF and we descended to 600 feet AGL above farmland at 400 KIAS. We started our low-level dialogue that was like the crew coordination used in an Intruder but different due to the automated system. We discussed the terrain clearance ahead of the aircraft and anything posing as a threat to our safe passage. We each had a radar display that showed similar but different, complimentary data.

At Passo Corese, we banked left toward the Fabrica De Roma train station, twenty nmi to the northwest. The system indicated a three-minute transit to the station. The terrain was flat providing little challenge to the TF system. Alberto completed a cockpit check on the hydraulics, engines, fuel and system. The aircraft was performing well and we had plenty of fuel. The system banked right above the train station and we flew three more minutes to the dam at the west end of Lake Corbara. A bird suddenly appeared near the aircraft but away from the engines. There was no indication that it hit us. As we approached the dam, we intentionally offset left for noise abatement. The dam was initially hidden by the surrounding terrain then began to show on radar.

We made a left turn to continue another three minutes to the expressway overpass at Ponticelli. We started flying over rolling hills and the TF responded superbly with the radar altimeter topping out slightly above 600 feet. We made a slight turn to the north after the overpass on our way to the town of Bettolle. Our route paralleled west of the A1 expressway over prominently flat terrain. We offset south at Bettolle for noise and made a left 90° turn southwest toward Torrenieri.

The terrain rose from 800 feet MSL at Bettolle to 2000 feet MSL in rolling hills then back down to 800 feet at Torrenieri. The TF system performed flawlessly as we continued our dialogue confirming terrain clearance. We turned south and traveled 12 nmi to Arcidosso. The terrain rose gradually along this leg to 2100 feet MSL. There was rapidly rising terrain to 4100 feet MSL in a 90° left turn with a dark radar screen at 1nmi. This indicated that we had nine seconds until terrain impact. The aircraft and TF system handled the mountain without a problem as Alberto spoke more quickly to verify clearance. Turning into rising terrain at high speed could be a lethal combination. The auto TF system reminded me of watching TV in my living room as the aircraft managed the terrain beautifully. This was an enhancing feature of the Tornado.

There were several higher ridges in front of us to fly over then the TF lowered us back into smoother terrain as our speed increased to 415KIAS. We crossed a valley and started our climb toward a tall radio tower a mile south of Ospedaletto in the Parco dei Sette Frati. I requested that Alberto direct the flightpath of the aircraft on a collision course with the antenna and not follow TF directions to determine the system response to the impending collision. As we approached the tower, the TF system directed a pullup then disconnected with the 'WOOP' audio warning and red lights indicating the danger ahead. Alberto manually flew the aircraft adjacent to the tower making a right 90° turn back down toward the dam at Lake Corbara.

We climbed up to a VFR cruising altitude and headed back to Pratica checking in with Rome Approach as we transited south for landing. We landed safely and debriefed the flight. I was very appreciative of this last flight as it provided an impressive demonstration of the low altitude, automated terrain following characteristics of the aircraft. I thanked Alberto for his tremendous help and made my way back to the hotel.

Tuesday 22 May I drove the rental car back to Rome airport and flew home commercially writing paragraphs along the way. I arrived at Dulles airport at 10pm and got home after midnight.

Wednesday 23 May The DTII writing continued in earnest. There was so much to do.

Thursday 24 May As several of our classmates were still away on their DT-II assessments, formal academic classes were suspended. Those of us that had finished the flights were busy writing our final reports.

Friday 25 May I had a T-2C flight with Randy in the morning and returned home to continue writing in the afternoon. I wasn't sure how it would all come together by the two-week deadline following our last flight. Hours upon hours of writing, proofreading, editing and photocopying were ahead of me.

Saturday 26 May I was busy writing paragraphs when the phone rang. I was surprised to hear Axel's voice. He was now the Chief Test Pilot for Strike at PAX. Axel asked, "Marble, what are you doing?" "I'm writing DT-II paragraphs", was my sad reply. "Come down to a party at the beach house on base", he responded joyfully. "I can't. I'm way behind. I have so much to write", came my response. "Marble...", Axel paused with a stern tone in his voice. "The work will still be there tomorrow. Bing is here on a cross country. We have made a big batch of chili. Put down the work and get over here", he continued adding a touch of authority to his words. "Yes, sir. I'll be right over", was my reply.

I didn't really want to go. I didn't think there was time to spare. I needed to keep writing. Reluctantly, I went over to the base to join Axel's party. His face lit up with his warming smile when I arrived. That made me feel great. The chili was delicious, the beer cold and the company the best. Going to the party was absolutely the best thing I could have done that day. It is important to make time for important events with friends and family. Tragically, this was the last social event that Axel and I shared and am grateful that he called me that day.

Monday 28 May Academic classes recommenced. It was difficult to concentrate on the content when we all knew there was report writing to do.

Thursday 31 May Today was my last jet flight at USNTPS. It was in a TA-4J with Air Force F-15 Eagle pilot and classmate, 1MIG. The A-4 was a rather simple aircraft compared to his Eagle but was still a great little jet for both of us. We considered our options for the flight, decided to stay low then go high. We headed down to the southern end of Restricted Area 4005 near Smith Point, Virginia and turned northerly while flying at the SOP minimum 200 feet above the water while flying by boats in the Chesapeake Bay. Boats on the Bay didn't get that view often as we were usually high up in the clouds. Once at the northern end of the restricted area, we pitched up into those clouds to fly among them in a low-level manner. We weaved our way up and down through the valleys and peaks created by them. This was one of my favorite pastimes. After chasing clouds for a while, 1MIG suggested practicing flight test maneuvers. He did a few wind-up turns, step input damping tests and an acceleration run. We came back for landing feeling joyful after a superb flight.

Monday 4 June I submitted my smooth DT-II Final Report within the deadline. It had been a challenging journey. The report had a total of 178 pages with 98 of that number being the weapon system description with many diagrams to add clarity. It was a great relief to have the report completed.

Friday 8 June An open cockpit, Pitts Special biplane, Registration number N9CC, had been at TPS for a few days providing flights for the pilots then a few lucky NFOs. The flights were in the morning and the classes in the afternoon. It was always easy to see which students had morning flights in the Pitts as their eyes were bloodshot. Part of the Pitts flight was an inverted spin that subjected the occupants to negative three Gs of load. That was a very large negative load that resulted in the blood of the aircrew rushing to their head resulting in blood-shot eyes. I was fortunate to be scheduled for the Pitts on my last flight at TPS.

The Pitts pilot, Pascarell, briefed the flight including what to expect in the inverted spin. Due to the open cockpit, he made a point of explaining the necessity of having my lap belt straps pulled tight for the flight. During the inverted spin, the aircraft would try to throw us violently out of the cockpit. I understood the point and made my lap belt tight. The flight was quite enjoyable and unusual with the open cockpit air blowing past us. We flew out to the training area and set ourselves up over open farmland.

Pascarell placed us in the inverted spin near 4000 feet MSL. I found myself hanging upside down from my lap-belt as the aircraft tried to throw me out while staring down at the rapidly closing terrain below. The inverted spin probably lasted only ten seconds. That was one of the most dramatic things I'd done in an aircraft to date. We landed back at PAX, I thanked Pascarell for the experience and went to class with bloodshot eyes. That was a wild ride! While the Pitts had an open cockpit and was lower to the ground, this video gives you an idea of what an inverted spin looks like[193].

Monday 11 June I received my graded Final Report. Crutch and Duncan[194] provided encouraging comments and specific areas for improvement. I appreciated the positive feedback as it had been a challenging assignment.

Saturday 16 June Many family members attended our graduation from USNTPS at the PAX Officers Club. All the graduates wore their dress uniforms. It was a great social engagement and celebration for finally completing a very difficult year of work and flying. While our class really enjoyed one another's company, it was time to move on and get real-world testing completed.

Prior to leaving the east coast, my new Commander Operational Test Force (COMOPTEVFOR) Operational Test Coordinator (OTC), Frenchie, requested my presence in Norfolk for a briefing on my new role with VX-5. COMOPTEVFOR was the coordinating agency for all Operational Testing conducted by the Navy. VX-5 served as the air-to-ground ordnance portion of that testing. Frenchie was a Navy Commander that had served as an A-6 BN and was serving as the OTC for classified programs. My clearance paperwork was processed to enable a briefing regarding classified developmental programs including the new A-12 aircraft.

---

[193] Inverted spin world record. Love the radio comms during the spin; https://www.aopa.org/news-and-media/all-news/2014/march/17/inverted-flat-spin-record-attempt

[194] General 'Duncan' Heinz became the Program Executive Officer (PEO) for the F-35 aircraft program prior to his retirement from the USMC in 2010. He had originally been an Intruder BN; http://www.metalsnews.com/t1118075i

I traveled to Norfolk and after identity verification, sat in a vault with no windows to be 'read into' classified programs. Frenchie pulled out paperwork regarding the A-12 aircraft and showed me a picture depicting the design; it was a dark, flying wing with tandem seating for a crew of two[195]. Frenchie and I briefly discussed the design, range, payload, carrier bring-back, top speed and other carrier aviation design essentials.

I left the vault vowing not to discuss the programs briefed except with other properly cleared and documented individuals in properly designated spaces. The next few years were going to be interesting. I started the drive to California looking forward to tall mountains and the wide-open spaces of the western USA.

---

[195] https://en.wikipedia.org/wiki/McDonnell_Douglas_A-12_Avenger_II. Aviators called the A-12 'The Flying Dorrito chip' as its shape resembled the snack.

# Chapter 11

## VX-5 Vampires, NAS China Lake, CA

*I was amazed at Bullet's flying, communication and tactical awareness. We had air to air threats, surface to air threats, a wingman to keep in position and terrain to avoid while pulling up to 5Gs at night at 500ft AGL. Bullet expertly managed it all and provided superb fighter support for the strike package with simulated missile launches on enemy combatants.*

VX-5 was the Operational Flight Test Squadron for Navy and USMC attack aircraft, weapons and their onboard systems. The squadron was one of many activities aboard NAS China Lake. The base is in the high desert, a three hour drive north-east from Los Angeles (LA) and an hour north of Edwards AFB. The NAS elevation is 2200 feet MSL with desert basins and mountains throughout the region. Mount Whitney, the highest mountain in the continental USA at 14,505 feet above sea level, is located 60nmi northwest of the base. Getting orders to VX-5 was a thrill for any aviator as they recruited the best personnel, had the best aircraft, equipment and airspace to test the newest weapon systems.

I arrived at VX-5 in mid-July 1990 to open arms and quickly became one of the Vampire family. The squadron enjoyed doing everything together. Being in a rather isolated desert community, we all bonded and made our own entertainment. On my first weekend at the squadron, a large group went up into the nearby Sierra Nevada mountains for a Chile Cookoff that we won! I was invited onto the winning team (Satch, Spud, Killer and I) and taught how to slice and dice the ingredients for the winning concoction. USMC Harrier pilot, callsign Satch, was the culinary director with the winning recipe. USMC Cobra pilot, callsign Spud, Intruder pilot, callsign Killer, and I provided the outdoor kitchen labor. Squadron families camped out in every imaginable tent, car, truck, trailer and mobile home and had a great time.

VX-5 flew the A-6E Intruder, EA-6B Prowler, F/A-18A/B/C/D Hornets, AV-8B Harrier, AH-1W Cobra helicopter, TA-4J Skyhawk and a Cessna 182 for logistics runs. I looked forward to getting back to flying the Intruder while continuing with the Hornets and Skyhawk. I was told that a flight in the Weapon System Officer (WSO) forward cockpit of the Cobra helicopter was a possibility as well. I looked forward to that experience.

Tuesday 24 July A-6E Intruder BN Charlie Braun and his pilot, Mines, were killed when dive bombing overwater at night. Charlie and I had been instructors together at VA-128. It was always painful to lose another squadron mate.

Thursday 2 August Iraq invaded neighboring Kuwait over various disputes including oil exports[196]. The invasion would be the focus of world attention for months to come. VX-5 aviators could see a war coming on the horizon. None of the crews that were at VX-5 in January 1991 saw combat in Desert Storm. We were on shore duty and excluded from the combat roles[197].

Monday 6 August A group of new VX-5 Operational Test Directors (OTD) were flown to Norfolk for a week of training at COMOPTEVFOR. The training provided background into how acquisition occurs, where money to test comes from and how we coordinate with a variety of agencies to accomplish testing[198]. We were given a manual to read cover-to-cover[199]. Despite the course being for OTDs, it seemed to provide more information on how to perform the role of our supervisors, the OTCs at COMOPTEVFOR. To be honest, the administrative structure behind flight test is daunting. As aircrew, we needed to focus on our part of testing the hardware and software without getting too lost in the complexity of the management. If interested, the reader is directed to seek further information about flight test management[200].

---

[196] This invasion changed global dynamics seemingly forever; https://en.wikipedia.org/wiki/Invasion_of_Kuwait

[197] As time passed, Desert Storm veterans joined VX-5 providing details of their experiences.

[198] The course; https://present5.com/comoptevfor-operational-test-director-s-course-navy-systems-acquisition/

[199] OTD manual; https://cupdf.com/document/operational-test-directors-manual.html?page=1

[200] https://www.acqnotes.com/Attachments/DoD%20Test%20and%20Evaluation%20Management%20Guide%20Dec%202012.pdf

Friday 10 August After the OTD class, I visited the Officers Club at NAS Oceana. Axel was there having dinner with his daughters. We smiled at one another, but I didn't interrupt his precious time. That was the last time we saw each other.

Tuesday 14 August My first flight at VX-5 was in an Intruder with Killer. He was a strong, brash pilot, fearless, enthusiastic and preferred to do things rather than talk about them. More action and less talking were just fine with Killer. He showed me the ropes of how VX-5 briefed flights. As this was a day warmup flight for me, Killer asked, "Do you want to go clockwise or counterclockwise around the restricted areas?" I didn't really care so he picked clockwise. We checked in with maintenance control and read the logbook prior to walking to our jet.

We completed a standard preflight inspection and then climbed into the cockpit. It was summer in the desert with the temperature was around 120°F. Everything in the cockpit was very hot. I clicked my Koch fittings into my parachute harness and could feel the heat from the metal attachment points burning through the flight suit. Killer explained that we wanted to get the engines started as soon as possible so we could close the canopy and get the air conditioning cooling things down. We followed that advice but the air conditioning in the jet seemed very inadequate. Once we received our takeoff clearance from tower and added full power, the air conditioning started working better but it took a few minutes before we started feeling comfortable.

After takeoff, Killer flew on a southwesterly heading at about 200 feet AGL. We switched frequency from China Lake Tower to a common range control frequency with Edwards AFB for traffic advisories. After five minutes, we arrived at the Honda automobile testing track about 35nmi southwest of the NAS. Killer explained how Honda tested cars on the track, but we used it as a visual reference for racing as well. He added maximum power and built up our speed to 480KGS. We entered the track on one end of the straightaway, spent 10 seconds to get to the far end, pulled 6.5Gs at the turn for about 15 seconds then another 10 seconds of straightaway before our final turn at 6.5Gs. We had covered the 7.5-mile track in 50 seconds[201]. Killer was a very good pilot and demonstrated that during that low altitude maneuvering.

We continued southwest for the next couple of minutes to the windmill farm west of Mojave airport then turned north passing Lake Isabella and eventually finding the Kern River Valley. I'd flown this valley while on my NVG hop during our TPS visit. While we could fly lower on NVGs than SRTC, we could get even lower flying visually on a clear day and did so. As the Kern River valley began to narrow, we made a turn to the east and were cruising along the top of a large plateau at 9000 feet MSL. It was quite beautiful with streams, ponds and open meadows. As we arrived at the edge of a steep drop-off, Killer rolled inverted and pulled to follow the contour of the terrain. He rolled upright into a 30-degree dive as our speed picked up to 450KIAS as we descended to the Owens Lakebed[202].

We resumed low altitude flight at 200 feet above the ground and continued north along the I-395 freeway. It must have been a thrill for those cars and trucks to see us screaming past them. Killer pointed out Mount Whitney on our left as we hugged the mountains across a valley to the east. We continued north until reaching Big Pine and Lake Tinemaha as it marked the northern boundary of our restricted area. We headed east up over a ridge of mountains south of Waukoba Mountain and into the Saline Valley. The scenery was stunning as we breezed effortlessly along. We passed by hot tubs that Killer mentioned were a good spot to go camping and then flew further south into the Panamint Valley. As we arrived at the southern end of the Panamint, we flew around a large grouping of rock spires called the Pinnacles near the desert town of Trona[203]. They were impressive to see. We headed back west, contacted China Lake Tower and came into the break before getting our clearance to land. We shut down the jet, headed inside to complete the paperwork and debrief the flight. We certainly had a great testing area to fly in! I looked forward to an amazing tour of duty.

---

[201] This was an exercise that we'd do many times while at VX-5 and even faster in a Hornet with 7.5G turns.
[202] This drop-off was a favorite spot for VX-5. We'd repeat the maneuver countless times in various aircraft.
[203] The Trona Pinnacles; https://www.blm.gov/visit/trona-pinnacles

Wednesday 15 August Test Pilot CAPT Steve 'Axel' Hazelrigg[204] was killed during a flight test in an A-6E Intruder flown from PAX[205]. His BN, 'Catfish' Davis, was badly injured but survived the event. Axel was serving as the Chief Test Pilot for the Strike Test squadron at the time of the accident. His aircraft suffered a catastrophic loss of the longitudinal control system.

It took several days before I heard of the accident. I was devastated by the news. He was my instructor pilot for my first Intruder visual low level, first 40-degree dive bombing and first VR1355 low level. Bing and I had spent memorable personal time with Axel's family. He encouraged me as a student in VA-128 and then again as a student at TPS. Axel was well loved throughout the Intruder and flight test communities. He was a naval aviation legend and the nicest guy you'd ever want to meet. Those of us who knew him well will forever mourn his loss. The large hangar planned for the A-12 aircraft at PAX was named after Axel. The hangar never saw use by the A-12 but was used later by the F/A-18E/F/G Super Hornet and F-35 programs.

Thursday 16 August My second flight at VX-5 was with Tom in an F/A-18D. He was a mild-mannered guy that didn't talk much yet was full of confidence. I suspect Tom was another genius in a flight suit. It was rather common in naval aviation and seemed more so in flight-test. We ran around the area at low level like Killer and I did in the Intruder.

Friday 17 August Craw and I flew an F/A-18D cross country to Buckley AFB Base in Denver, Colorado. This was fitting on many levels. He and I had both had studied together and graduated in aerospace engineering from CU Boulder, had served together in NROTC, then served together aboard the USS Midway.

Once back at VX-5, I started learning about the projects that were in work. The A-6E guys were testing the new mission computer software E250 with all its improvements along with the AGM-123 Skipper missile. They were getting ready to fly NVGs once we could get an aircraft with lighting modifications. They were participating in the

---

[204] Memories for Axel;
https://usnamemorialhall.org/index.php/STEVEN_A._HAZELRIGG,_CAPT,_USN
[205] The accident details; https://aviation-safety.net/wikibase/57203

Intruder Block 1A upgrade evaluation. The Hornet guys were testing the new 91C mission computer software upgrade and evaluating the Infrared Maverick, Harm Block IV and Standoff Land Attack Missile (SLAM) missiles integration with the aircraft.

I was assigned to the VX-5 Future Systems Office (FSO) upon arrival. The FSO provided OTD management and supervision for the A-12 Avenger aircraft program and other highly classified programs. Serving as an OTD in the FSO involved a lot of classified reading and writing while attending many classified meetings. The work was very interesting but the classification of the programs was burdensome. Our security specialist, Tammy, ensured that all classified documents were managed meticulously.

I was given a thorough, face-to-face FSO turnover with outgoing Intruder BN OTD, callsign Scanner. We traveled together to various clandestine locations to attend highly classified meetings attended by contractor and military personnel from all over the country. At times, we wore civilian clothes to the meetings to further disguise our military affiliation. Throughout our travel, Scanner introduced a cadre of participants involved in a variety of projects. After several weeks of travel, introductions and turnover, Scanner headed to Whidbey Island to rejoin the Intruder fleet.

I served as the lead OTD on a variety of FSO projects and Assistant OTD for the A-12 aircraft. My office mate, F/A-18 test pilot, callsign Decoy, served as the lead OTD for the A-12. Some of the programs we worked on proved fruitful with effective products for naval aviation. The A-12 and others did not. We wished that circumstances allowed us to help fix the troubled programs but that was not our role. We were to determine if the weapon systems were operationally effective and suitable for naval aviation. By the time the product arrived at VX-5 for testing, design and manufacturing problems should have been eliminated to the best of the government and manufacturer's ability. This was often easier said than done.

Every aviator brings his/her own flying experience to work each day. Our perspective and decision making are influenced by significant events in our flying careers. Decoy and I shared the USS Midway during our initial junior officer tours. Despite being on the same ship, our perspectives on flying differed greatly in many ways. Decoy excelled flying a single seat aircraft with technology that was state of the art for the time. I served in a two-man crew in an aircraft that was designed in the 50's and slowly upgraded over time. Despite its age, the Intruder could fly night missions close to the ground to defeat enemy weapon systems while pursuing unlit targets that could require dedicated radar interpretation and targeting. This mission required the two-person crew to work well together to complete the task safely and successfully. The Intruder crew coordination and unique technology enabled us to complete missions that were simply not feasible with the F/A-18. This comment is not intended to slight the Hornet community or their crews. The aircraft were simply designed to do different things. The A-12 was designed to successfully accomplish the Intruder mission taskings.

For many casual observers, the joining of a Hornet pilot and an Intruder BN to work collaboratively on the new A-12 aircraft made great sense. The Navy could take the best insights of both communities and allow that to flourish in the new A-12 aircraft. The A-12 cockpit design mirrored the Hornet cockpit as they were both built by McDonnell Douglas. Decoy's experience with the Hornet was invaluable in evaluating the new design. I didn't have his extensive experience in Air-To-Air (A/A) engagements or with a newer cockpit. Decoy had arrived at VX-5 six months before me (he was in Class 96 at USNTPS) so he was ahead in understanding the program, players, the aircraft and various developmental concerns.

3 - 8 September Craw, Navy flight surgeon, Hornet pilot and NVG guru, Doc Antonio, Hornet pilots Bullet, C.T. and I traveled to London to attend a NVG Symposium and the Farnborough Airshow[206]. Our primary reason for our attendance was to examine the latest NVG technology along with discussions on future needs for a Helmet Mounted Display (HMD). We were disappointed at the speed at which the HMD technology was improving[207].

---

[206] An impressive airshow;
http://www.globalaviationresource.com/reports/2011/fia90.php
[207] The Joint Helmet Mounted Cuing System (JHMCS) was eventually developed in response to an identified need. It revolutionized combat for F/A-18, F-15 and F-16

At Farnborough, we all wanted to get a look at the Soviet MIG-29 and watch its 'Cobra maneuver'. The maneuver highlighted the aircraft's great agility with high angles of attack and amazing pitch control while flying close to the ground. The maneuver was beyond anything a US aircraft could accomplish at that time and led to concern about the superiority of the modern Soviet fighters. Watching the MIG-29 fly was an eye-opener with its impressive demonstration of maneuverability. Alex Zuyev had defected to Turkey in May 1989, but I hadn't read any intelligence reports that clarified the aircraft's capabilities.

October Pink Pig flights. While testing the IR Maverick missile integration on the Hornet, a commercial advertising company decided to build a huge pig in an isolated area of the desert within a ten-minute flight south of China Lake. The pig was about 50 feet tall and 100 feet long. It was supposed to represent a piggybank using the notion that you could save money if you purchased the advertised car as its running costs were so low. Naturally, this pig served as an excellent target for our simulated weapon releases. We conducted 'Pig Runs' which meant that we executed simulated attack profiles without a weapon release. It would have cost the Navy a great amount of money to construct such a great target in that location but we got to use it for free. Sadly, once the advertising campaign was completed, the pig was dismantled and removed. As aircrew, we were sad to see it go. It had been a wonderful part of flying around China Lake.

Wednesday 3 October Boomer Intruder pilot, Dish, became a Vampire after his fleet tour. He gave me an annual NATOPS check ride in the A-6E.

Early December VX-5 received word that Soviet built hardware had been recently relocated to West Germany and was available for testing. As the hardware involved was also used by the Iraqi military, their operational capabilities were of interest to our aircrews. Killer, Hornet pilot Slider, intelligence officer, Mark and I went over to Germany to check out the hardware. To test the Soviet system, we needed U.S. fleet aircraft. We borrowed a couple from a carrier on deployment in

aircraft. Its tactical advantage in the F/A-18F aircraft is recognized in Brad Elward's book. They are realizing our initial dream (Elward, 2012, pp. 165-167).

the Mediterranean. Our VX-5 team flew commercially to Germany, met the fleet aircrew that had delivered the Intruder and Hornet aircraft and the operators of the Soviet equipment.

We traveled to a remote location where the Soviet hardware and East German operators were located. We examined the hardware and asked questions about its strengths and weaknesses. We had a good exchange of information with the help of a translator as we didn't speak German. We thanked them for their assistance and looked forward to flight testing the next day.

We briefed the missions to test the hardware. I flew with the fleet Intruder pilot from the carrier. This allowed me to analyze the Soviet systems in real time. The fleet Hornet pilot flew his aircraft. I briefed him on what to test and how. We mapped out the flight paths and maneuvers that we would execute. Data was recorded and debriefed after each flight. After five days, we flew back to China Lake and got to work collating the data. Priority Secret messages were sent to update the fleet on our findings[208]. Secret articles were distributed in the Bat Bulletin[209] to better describe the results.

Monday 7 January 1991 The A-12 aircraft acquisition program was cancelled by Secretary of Defense, Dick Cheney. The Intruder community was devastated as they counted on the new aircraft to replace the Intruder yet retain the experienced, all weather attack Navy aircrews that provided a potent warfighting capability for the United States. Intruder aircraft upgrades had been deferred to secure the A-12. Legal battles over the program's cancellation went on for a decade with great amounts of money spent without the delivery of a new aircraft. The program provided lessons learned in acquisition for the government and contractor alike.

As the A-12 was cancelled, Decoy shifted his attention over to other projects, most importantly the final Operational Evaluation

---

[208] Within a month, military aircrews were facing the systems that we tested but in combat with Iraqi forces.
[209] The Bat Bulletin was a secret, quarterly publication published by VX-5 and distributed to Navy and USMC tactical squadrons and various intelligence organisations.

(OPEVAL) for the radar jamming hardware, AN/ALQ-165 Airborne Self Protection Jammer (ASPJ). OPEVAL was the final testing with production representative hardware and software before the DOD decision to purchase the system and deploy it to the fleet.

Decoy excelled as the OTD for the ASPJ project. His team collected flight test data to determine if the radar jamming hardware helped protect aircraft. The test team then used a statistical technique called Analysis of Variants (ANOVA) to determine system effectiveness. They discovered that the ASPJ hardware did not provide substantially better performance than did an aircraft without the hardware. As a result, the ASPJ hardware failed OPEVAL. Decoy traveled all over the country to explain the failure to high level military and civilian leaders[210].

Thursday 17 January Desert Storm[211] started with F117 stealth aircraft bombings followed by a barrage of international firepower. The world watched unprecedented media coverage that brought the war into lounge rooms around the world. U.S. Army General Schwarzkopf emerged as a strong, charismatic leader that without question was going to push Iraqi troops out of Kuwait.

Friday 18 January Hollywood released the movie, 'Flight of the Intruder'. The timing was interesting considering the active Intruder combat operations in Desert Storm. I enjoyed watching the movie on a military base with a military audience. It was clear, however, that this movie was no box office smash like 'Top Gun'.

Sunday 20 January Intruder pilot, Jake and I flew one of our A-6Es from China Lake to NAS Fallon in 45 minutes. The Vampires deployed fourteen aircraft and three hundred personnel to Fallon to test and develop tactics with the Block Twelve, Night Attack (NA) Hornet. We also hoped to support Desert Storm in the event naval leadership wanted to send any of our crews as subject matter experts. The powers in

---

[210] This time provided Decoy a strong foundation for his career in flight test and acquisition.
[211] Following Iraq's invasion of Kuwait, a multinational force assembled to stop further transgressions. This effort was named Desert Shield. Diplomacy did not produce an Iraq withdrawal leading to forced conflict in operation Desert Storm; https://www.history.com/news/history-vault-operation-desert-storm

charge of the Desert Storm air war were confident they had everything under control without our help. The air-plan had been laid out in detail and everyone was ready to execute the plan.

Tuesday 22 January Hornet pilot, Bullet[212] and I flew a night NVG fighter cover mission in restricted area R-4816N/S[213] while carrying a TACTS pod. The pod provided real time aircraft tracking for scenario evaluators and for replaying the events at the mission debrief. The restricted area had a valley at 4000ft MSL in the center running 020°/200° True and mountains on either side rising to 8850ft MSL. The vertical boundaries of the areas were 500ft AGL to FL180. In the preflight brief we discussed the moon being 38% of full and an hour past its highest point in the sky for our takeoff at 630pm. That was sufficient illuminance for the goggles to perform well.

We manned the jet and flew in formation with our wingman out to the restricted area. Once on station, we broke the formation into single aircraft with a 2nmi separation. We kept our wingman in sight with the NVGs. We waited for all the mission players to arrange themselves into position before the exercise commenced. I was amazed at Bullet's flying, communication and tactical awareness. We had air to air threats, surface to air threats, a wingman to keep in position and terrain to avoid while pulling up to 5Gs at night at 500ft AGL. Bullet expertly managed it all and provided superb fighter support for the strike package with simulated missile launches on enemy combatants. This was a tactical application of the NVG capability experienced in the Kern River Valley while on my TPS visit to China Lake.

Wednesday 23 January Bullet and I had a repeat of the mission from the previous night to enhance our proficiency. This helped reinforce coordinated tactics and strategies using NVGs. The moon was a bit brighter at 49% illuminance. Bullet was great in explaining things as we flew. Combat operations using NVGs was planned for the A-12 but something else needed to fill the gap following the cancellation of the program. Looking through the narrow field of view of the goggles while at low altitude with a dynamic tactical situation, caused our heads to be constantly moving, looking here then there. There was a lot to take in to

---

[212] Bullet was an A-7E pilot aboard USS Midway. He excelled flying Hornets.
[213] Fallon map; https://steadusers.org/wp-content/uploads/2018/12/Fallon.jpg

ensure safety as well as engaging the opposition as fighter cover. The air-to-air story was on the glowing green cockpit displays in front of us. It would have been easy to focus on them, but we were low to the ground. We had to keep a continual sense of the terrain around us. This new concept of low altitude, NVG fighter combat was thrilling.

Thursday 24 – Friday 25 January I had two Intruder flights, one with pilot Zot and another with Dish. We conducted strikes to one of the Fallon bombing ranges and dropped MK76 blue bombs. We were opposed by F-15 Eagle aircraft flying out of Nellis AFB near Las Vegas. It was good to have our Hornets to keep the Eagles busy so that we could drop our ordnance.

Monday 28 January Hornet pilot Zoil and I had a day flight where we served as a simulated HARM shooter and fighter cover for a strike package. He was the OTD for the HARM Block IV missile that was undergoing testing. Zoil knew the missile inside and out and was an enthusiastic teacher for all things HARM. It was a great experience flying with him. Zoil was a down-to-earth guy that didn't pull any punches when telling folks exactly how he felt. That came very handy in Block IV testing when the contractor tried to get Zoil to test a missile that wasn't production representative. He wouldn't have any part of it and stopped testing awaiting the contractor to come up with production representative hardware.

Tuesday 29 January Jake and I flew a night mission to the local bombing range. He was a big guy with a teddy bear disposition unless provoked. Jake had a dry sense of humor and was a gifted Intruder pilot. It was always a pleasure flying with him[214].

Wednesday 30 January As our detachment to Fallon was ending, Craw and I flew another NVG fighter cover mission for an opposed strike. Like the prior missions with Bullet, we were in a flight of two aircraft at 500 feet AGL supporting our attacking forces. It was a pleasure to watch him flying aggressively near the ground, keeping his wingman in sight, prosecuting the inbound bandits and keeping us safe. Craw and

---

[214] Bruce's father, Admiral Lyle Bull, was a legendary Vietnam war Intruder BN.

our other pilots represented the incredible strength of our naval aviation forces[215].

11 - 29 March I completed a modified Cat IV Hornet WSO syllabus with the USMC west-coast FRS, the VFMAT-101 Sharpshooters at MCAS El Toro, Irvine, CA. Despite flying the Hornet, it was essential to better understand the aircraft and how to use it with future classified programs. I completed Operational Flight Trainer (OFT), Weapons Tactics Trainer (WTT) simulator and Part Task Trainers (PTT) training at the FRS.

I had the pleasure of flying in the Hornet with a variety of Vampire pilots and was impressed by all of them. Decoy and I only flew four flights together while at VX-5. They were day flights in F/A-18D aircraft in April and May 1991. I recall our first air-to-air flight with adversary aircraft. Decoy controlled the radar, had great situation awareness, tactical prowess and maneuvered the aircraft skillfully at 7.5Gs to gain advantage over our opposition. Maintaining physical fitness was a priority for Decoy and it showed during demanding maneuvering. He would be a pilot of choice to protect your aircraft in combat.

I was approached by a mathematician and analyst, Alice, from the Naval Warfare Center (NWC) China Lake. NWC had received Chief of Naval Operations (CNO) tasking to debrief combat crews from Desert Storm. Alice asked if I was willing to travel around the country debriefing crews then helping to document the results. It appeared to be a great opportunity to see what worked in combat and what didn't. Alice and I interviewed F/A-18 Hornet, A-7E Corsair and A-6E Intruder crews at NAS Jacksonville, FL, MCAS Beaufort, SC and NAS Oceana in Virginia Beach, VA. The aircrew were more than happy to discuss their experiences, the things that worked well and the problems they faced.

Monday 20 May Dish and I shared my first NVG flight in an Intruder. The original Intruder cockpit had red night lighting throughout the cockpit. The aircraft had to be modified to exchange all that red lighting for green that would not produce blinding glare under NVGs. We briefed the introductory flight as a counterclockwise low-level route in

---

[215] Despite his tactical skill, Craw started flying commercially after this detachment.

the restricted area. We took off at 9pm with a 50% full moon with its highest point in the sky around 7pm. That provided plenty of moonlight for the NVGs to provide a clear scene.

After takeoff and once clear of the tower environment, we donned the goggles and proceeded down to low level. While I preferred being in a front seat in the Intruder, the bulletproof windscreen made the NVGs less effective. The windscreen appeared tinted resulting in a more obscure view than looking out the side of the aircraft. As with the F/A-18, flying on NVGs turned night into day. We flew at 200 feet AGL throughout flat and rugged terrain and felt comfortable doing so. NVGs provided a new advantage to tactical aviation.

Tuesday 21 May Hornet pilot, callsign CT, and I shared another NVG flight in an F/A-18D. He was a Navy Commander and the squadron Chief Operational Test Director (COTD). He was a very bright, easy-going man with a quick, potentially cutting, sense of humor. He looked after the JOs during the flight test chaos of our work.

Like the night before, CT and I took off at 9pm to ensure that it was sufficiently dark. The moon was over 50% full with its highest point in the sky at 8pm. After takeoff and clear to the east, we donned the NVGs. We settled onto the low level and were comfortably flying at 200 feet AGL up the Panamint Valley. Due to the bright moon, it felt like day. We crossed over a ridge line near Panamint Springs into the Owens dry lakebed and continued north. As we approached the Tinemaha Dam, we turned east and entered the Saline Valley and proceeded south back to China Lake.

On arrival back, we asked the tower to turn off the runway lights. They cleared us to land as we were the only aircraft flying. Landing while using NVGs was authorized under specified conditions. The landing proceeded like one during the day.

Tuesday 16 July While at VX-5, I became concerned with our naval aviation acquisition system. I'd witnessed the cancellation of the A-12 advanced strike aircraft, the poor OPEVAL performance of ASPJ and serious developmental delays with many other programs. I felt the fleet deserved better. In a hallway of VX-5, there was a large sign; 'What have you done for the Fleet today?' In contemplating where I could make a greater impact, I applied for a transfer to the Aerospace Engineering Duty Officer (AEDO) community with my CO's encouragement. Due to their insight into acquisition and flight test experience, many AEDOs were military test pilot school graduates. This change in career designator took me away from the Intruder community. I hoped to deliver better products to the fleet in a more expeditious manner[216].

Friday 26 July USMC pilot Spud and I shared a day flight in the AH-1W Super Cobra attack helicopter[217]. He was a thin, athletic and quiet Marine who enjoyed time with his mates but had a no-nonsense attitude when it came to aircraft development. Spud was honest to the core and wouldn't pull any punches when evaluating a system. I looked forward to flying with him and sit up front in a tandem configured cockpit. The WSO's station was the front seat and the pilot sat behind the gunner.

The field of view out the WSO's front and side canopy glass was excellent. Spud started up the engines, we lifted off and were cleared to the west. We stayed over the unpopulated, flat desert terrain as we flew toward Highway 395 at 50 feet AGL at 120KGS[218]. We climbed into the mountains near Kennedy Meadows where our squadron had won the Chili Cookoff the prior July. We continued north into the higher mountains then slid down into the dry Owens Lakebed. We made a few simulated attacks on old mining hardware at the lake then headed home past the Haiwee reservoirs and south down 395. Spud had a sweet touch as a pilot. He suggested we refly the mission on NVGs. I looked forward to the opportunity.

I've flown with many amazing pilots that have treated me to incredible vistas while flying. One pilot I never flew with but feel especially blessed to have met is Soviet Air Force MIG-29 fighter pilot and defector, Alex Zuyev. He was an incredibly brave patriot of the Soviet Union that powerfully acted then spoke out against injustice in his country of origin. We worked very briefly together on his book. I

---

[216] In the years that followed, I learned that military material acquisition is challenging without simple solutions.
[217] The AH-1W; https://en.wikipedia.org/wiki/Bell_AH-1_SuperCobra
[218] A gunner's seat video; https://www.youtube.com/watch?v=SztpM5Z6JLM

appreciate that most people may have no knowledge of Alex Zuyev. I suspect the highly regarded journalist, Mike Wallace, wanted everyone to know his courage, intellect and sense of humor as he interviewed Alex on the 60 Minutes program in January 1993[219]. Alex's story is one that Russian government officials from the past and present would prefer is never heard. I highly recommend reading his book, *Fulcrum*.

Alex was a loyal pilot of the Soviet Union that excelled in his flying ability. His performance earned him the right to fly one of the Soviet's most advanced aircraft, the MIG-29. During his career, his mother suffered medical problems that the doctors were unable to remedy. Alex's mother became a vocal critic of the Soviet medical system. The Soviet leadership did not appreciate her criticism and placed her in an insane asylum. Her incarceration opened Alex's eyes to the truth of Soviet society.

Alex discovered that the South Korean airliner KAL 007 was shot down by a Soviet SU-15 fighter jet on 1 September 1983 killing all 269 people onboard. The reason for the shootdown was to save face for the Soviets due to maintenance issues with their ground-based radar systems. The Chernobyl nuclear accident in 1986 resulted in needless lives being lost due to poor and selfish leadership decisions. On April 9, 1989, Soviet military troops attacked civilians in Tiblisi to control uprisings resulting in the death of twenty-one civilians. Alex and his squadron mates were told that they may be required to use their aircraft to stop the uprisings. Alex acquired a contraband radio and listened to Western news and music across the Iron Curtain[220]. Alex saw the truth of the world and the lies of the Soviet leadership. He was motivated to escape from the Soviet Union.

Alex developed an elaborate plan, drugged his squadron mates with a tainted cake and stole a premier MIG-29 fighter aircraft on 20 May 1989. He flew it to Turkey and requested asylum with the United States. Alex accepted great risk in the process as he would have been killed if the plan failed. Turkey allowed the Soviets to retrieve their MIG-29 but the USA provided asylum to Alex.

---

[219] 60 Minutes with Mike and Alex; https://www.dailymotion.com/video/x2pn4w3
[220] Radio; https://www.wildworldofhistory.com/blog/gorbachev-and-rock-and-roll

Alex was a priceless asset from an intelligence standpoint. The West never had an opportunity to debrief a Soviet fighter pilot with his insight and skills. There was great interest in the MIG-29 and SU-31 aircraft at the time due to their amazing flight capabilities, lethality and secrecy. Alex was interviewed by military intelligence and the Central Intelligence Agency (CIA). Details of his interview were released in secret messages and sent out to the tactical military community to modify their training and tactics. I was stationed at VX-5 when the messages of Alex's debrief arrived. He and aircrew from TOPGUN started a tour to tactical squadrons around the USA.

Prior to their arrival at VX-5, our intelligence officer, Kevin, asked us to visit the squadron classified library and read the secret messages. He wanted us well prepared for the visit. That was a good idea but in practice didn't work so well. We were all busy with our projects. There always seemed to be something better to do than read message traffic. I felt that way but forced myself to the vault to start reading the day before Alex's visit in mid 1991. I found the reading interesting and developed questions for Alex.

U.S. trained aviators considered the MIG-29 to be a potent adversary in combat. My 1960's vintage A-6E would have been a rather easy prey for a competent pilot in the much more maneuverable MIG-29. Alex gave a closed-door brief to our aircrew that covered a portion of the items in the secret messages. When finished speaking, Alex opened the floor to questions. Unfortunately, our aircrew were left flat-footed as few had a chance to read the messages. I asked a few questions that seemed to resonate with Alex. He appreciated the questions and noted that I had done homework prior to his arrival. After all the questions finished, we thanked Alex for his time and effort in educating us on Soviet hardware and tactics. It was a pleasure to meet Alex. I never expected to see him again but was wrong.

Several months later, Alex and I were at a professional conference. He greeted me recalling his visit to VX-5 and my questions. I was surprised that he recalled meeting as he had just finished traveling across the whole country and briefed thousands of aircrew. Alex mentioned that he was writing a book about his escape from the Soviet

Union. He asked if I would read portions of the draft and provide feedback. Naturally, I was honored and agreed to his request.

Alex retrieved the manuscript from his room and joined me in the lobby of our hotel. I read his words as he sat quietly and intently beside me. When I found a section that didn't explain the events in a logical and simple manner, I suggested a change and provided substitute words. Alex was appreciative of my changes as he was still learning how to best communicate his story in English. We spent a couple hours together reviewing the manuscript. We both knew the sky, had disciplined military training and knew our roles in the Cold War. Yet when we spoke, Alex and I were away from our adversarial roles and got along well. We met at a more basic level of humanity and enjoyed one another's company. Alex thanked me for my assistance. I commented that the manuscript was enjoyable to read and looked forward to reading the final product. We said goodbye, never to see each other again. In the mid 1990's, an aircrew friend mentioned that Alex's book had been published and I was mentioned in the acknowledgements. I was honored!

While preparing a speech for my retirement from the Navy in 2003, I thought of the significant events and people of my career. Alex came to mind despite the short duration of our acquaintance. I did a Google search for him to sadly find that he died in an aircraft accident on 10 June 2001. The accident was attributed to an accelerated stall at low altitude resulting in a flat spin while Alex was serving as a flight instructor in the front seat of a Yak-52. The FAA flight incident report can be found here[221].

After reading Alex's book, I knew him much better and understood his path in life. Alex was willing to stand up to his perception of evil in his country of origin. He chose to do something very dramatic in the name of truth despite facing grave risk. Alex's actions were obviously seen as mutiny and a rejection of his country's methods of management. Soviet officials would naturally be quite upset with Alex and would rather he not live for very long. While it is possible that Alex or the commercial pilot in the aft seat stalled and crashed the aircraft, I

---

[221]https://www.ntsb.gov/_layouts/ntsb.aviation/brief2.aspx?ev_id=20010614X01180&ntsbno=SEA01LA116&akey=1

don't discount other possibilities including sabotage. Commercial pilots are well trained to avoid and recover from stalls and spins in aircraft.

August Dish, Skyler and I climbed Mount Whitney with a summit at 14,505 feet (4421m) above sea level. The climb included a hike of 22 miles (35km) roundtrip with an elevation gain of 6145 feet (1873m)[222]. The journey was an all-day event starting with a 5am departure from China Lake. As we were driving from the desert floor at 2500 feet elevation to the car park at the Whitney Portals at 8374 feet (2552m) above sea level, we noticed a senior citizen riding his bike up the mountain. We were impressed with his performance due to the steep grade of the road. We started the climb from the Portals at 7am.

Around 9am, the man previously riding the bike ran past us as we were trudging along the path and disappeared over the horizon. We were again very impressed as we were keeping a reasonable pace. Near 1pm, we were quite tired and oxygen deprived while enduring steep trail switchbacks at 12,000 feet of elevation. Our super-fit senior citizen was running down the mountain towards us. Hypoxia was controlling our ascent and this guy was still running. As he approached, I noticed his 'Los Angeles Marathon' T-shirt. I stopped him to hear his story. I asked, "Do you do this to get in shape for the Los Angeles marathon?" He replied, "No, I do the Los Angeles Marathon to get ready for this. I run up to the Mt Whitney summit about thirty times each summer". We were all stunned. He had the best aerobic fitness I've ever seen. If in doubt, try it. Good luck.

Skyler stopped on the switchbacks as he was not feeling well. Dish and I made it to the summit around 230pm. The view was glorious. We put on our jackets as the temperature was cool despite the blazing sun and clear sky. A glider soared above producing sounds of wind rushing off its wings as it flew by. The trip down was easier and we arrived at the Portals near 8pm for the drive back to China Lake. That was a long and amazing day!

Ensign Eric 'Bone' Hamm arrived at the VX-5 Vampires. He was a newly winged naval aviator awaiting his slot at VA-42, the east coast

---

[222] The hike is closed in winter; https://en.wikipedia.org/wiki/Mount_Whitney_Trail

Intruder FRS at NAS Oceana. Prior to joining the navy, Bone graduated with a degree in Physics and worked for the Saturn automobile company. He decided that life could be more exciting, so he applied to navy Officer Candidate School (OCS) hoping to fly. He finished OCS and jet pilot flight training before reporting to China Lake. Bone just finished flying A-4 Skyhawks in the training command and flew them at VX-5. I was fortunate to fly and share adventures with him during his short stay in the high desert.

Tuesday 10 September Bone and I had had flights in the TA-4J on the 10th and 20th of September. For our first flight, I suggested a low-level flying north up the Kern River Valley, a right turn to head west over towards Olancha Peak then dive bombing runs at the Owens dry lakebed. This was something that I'd done frequently with other Vampire pilots and thought it was a good idea to share the joy of that experience with Bone.

We tucked the Skyhawk down inside the Kern River Valley and progressed north. After five minutes in the Kern Valley, we took the right turn and cruised along the high meadows plateau east of Red Hill. The meadows are at 9000 feet above sea level and beautifully green with streams flowing through them. We flew over to the big drop-off at 9043 feet of elevation with the valley floor below at 3973 feet. As we passed the ledge, we rolled inverted and tracked down the mountain. We rolled upright into a 25-degree dive toward the Owens Lake. The journey down from the plateau only took 20 seconds while accelerating toward 450KIAS.

Once we arrived above the Owens Lake, we climbed back up to 14,000 feet to set up a 40-degree dive bomb pattern. There was a large steel tower in the middle of the lake that we used for simulated bombing runs. We set up our aircraft in a circular orbit around the tower, pulled the nose of the aircraft around and rolled inverted toward the target. We rolled wings level upright as we dove at 40 degrees nose-down toward the tower. We simulated a weapon release then a 6G pull stopped our descent for the climb back up to 14,000 feet. We did a couple of these 40-degree dive bomb patterns then returned for landing with a better appreciation for one another.

Friday 20 September Bone and I shared our second flight in the TA-4J. He was quite comfortable with the training area and the aircraft. Bone demonstrated great skill in maneuvering the aircraft on another low level around the restricted area. He was going to make a great Intruder pilot.

Thursday 10 October Intruder Pilot Dan 'Dewey' DeWispelaere[223] and his BN Grady Hackwith died while on a low-level training flight from NAS Whidbey Island. Dewey and I had been instructors together at VA-128. He was a great man and his death was another tragic loss for all who knew him.

Thursday 17 October The stars and moon finally aligned for my NVG flight with Spud in the Cobra helicopter. We had 2/3 of a full moon that rose at 3pm and didn't set until 1am. The sun set just after 6pm so we waited until after 8pm for a dark takeoff. Spud started the Cobra without goggles but we put them on prior to taxi. As we didn't have to eject out of a Cobra, we didn't have the concern to remove them before egress. The goggles had a narrow in field of view but turned the night into day as we taxied then lifted off higher to depart the tower pattern. We headed west flying at 50 feet. We retraced our steps from our first flight and headed to Kennedy Meadows. As we took off late, we avoided residual effects from the sunset as we flew west. The moon was high in the sky and provided a clear view with the goggles. The cockpit lighting was well integrated for the goggles with minimal interference.

I felt very fortunate to be experiencing this amazing aircraft, technology and Spud's piloting skills. As we approached the mountains and started our climb, Spud suggested I keep my eyes open for moving targets that might be bears. I kept my scan going from side to side but didn't see any movement. We arrived at Kennedy Meadows, flew around the elevated terrain and returned to China Lake. The flight was an amazing experience!

---

[223] Another great loss;
https://usnamemorialhall.org/index.php/DAN_D._DEWISPELAERE,_LCDR,_USN

Monday 28 October Intruder pilot, callsign PK[224], provided my annual NATOPS check ride in an Intruder. We blended the flight with practice bombing and missile testing. We headed over to Superior Valley and dropped six MK-76 practice bombs using various deliveries including pilot to BN FLIR handoffs with good hits. We carried HARM and IR Maverick missiles and developed section tactics with a squadron Hornet pilot.

In November the Vampires got ready to take Maverick and HARM missiles to the aircraft carrier. Jake, PK, E-2C transitioned to Intruder pilot, Chet and I had flights carrying the missiles on day flights to verify functionality and identify any anomalies in performance. The carrier operations were designed to shake the missiles before firing them in the months to come. We needed to ensure that failures didn't occur with the loads, vibrations and shocks of routine carrier flight operations. In the fleet, missiles can be loaded on aircraft for an extended period without the need to fire them. Exceptional reliability was required despite the environmental stresses.

Friday 22 November VA-115 pilot turned VX-5 pilot, Joe and I flew an Intruder loaded with IR Maverick and HARM missiles to the USS Ranger[225]. The carrier was located off the California coast. We made many arrested landings and catapults during 3.6 hours of flight time. Both missiles remained operational throughout the testing. Joe and I spent the night aboard the carrier and returned to China Lake the next day. The missiles remained fully functional and were ready to fire when the opportunity arose.

Tuesday 26 November Chet and I flew a four-hour cross country to Fallon to drop a rack of MK-76 bombs and test the IR Maverick and HARM missiles. Despite the carrier shaking, the missiles performed superbly.

December The Navy awarded concept exploration contracts to five aerospace companies to present proposals for a new tactical aircraft,

---

[224] PK is an enthusiastic individual with an engaging laugh. He was a fleet SLAM missile expert.

[225] The USS Ranger had been one of six active carriers during Desert Storm; https://www.navysite.de/cvn/cv61.htm

designated AX[226], to replace the Navy A-6 and the Air Force F-111 aircraft. The lead vendors for the five contractor teams were Grumman, Lockheed, McDonnell Douglas, Rockwell and General Dynamics. A recent USNTPS graduate and Intruder pilot, callsign Wingnut and I worked on the AX program at VX-5. As Intruder crews, we appreciated the painful loss with the cancellation of the A-12 program. We hoped to contribute and be engaged at the start of the AX development.

I took leave over the Christmas holidays to visit friends and family in Australia. I made my way down to Nowra, New South Wales (NSW) to visit with my TPS classmate, Miz, who was serving as a helicopter test pilot in the Royal Australian Navy (RAN). I enjoyed visiting the local Fleet Air Arm aviation museum[227] and driving west into the Snowy Mountains. I traveled north to see my sister, Janet, who lived in Northern NSW. VX-5 squadron mate, callsign Fish, flew to Australia and we enjoyed a road trip along the east coast.

---

[226] The A-12 cancellation led to the AX;
https://www.globalsecurity.org/military/systems/aircraft/a-x.htm
[227] An excellent military aviation museum is in Nowra;
https://www.navy.gov.au/heritage/museums/fleet-air-arm-museum

# Chapter 12

## Marriage, AX briefs and ACOTD[228]

*An impressive crowd estimated at 5000 were present for the decommissioning ceremony of the USS Midway. The Secretary of the Navy Garrett, 15 former Midway Commanding Officers and 25 original Midway crew (plankowners) from 1945 along with representatives from Japan were present. USS Midway had been in service for 47 years changing the lives of many people including my own.*

Sunday 5 January I arrived back to China Lake from Australia at 11pm. That is a long journey.

Monday 6 January I was quite tired from all the travel but made it into work. It was quiet in the office with many people still on leave. I left early in the afternoon with jetlag.

Tuesday 7 January Wingnut and I shared a 1.5-hour Intruder flight. We flew low level around the area as a warmup flight for me.

Thursday 9 January Chet and I flew north to B-17 at NAS Fallon, dropped twelve MK-76 bombs and returned to China Lake with two hours of night flight time. That was great training.

Friday 10 January Bone and I drove three hours to Mammoth ski area arriving at the VX-5 ski lodge at 1am Saturday. We skied Saturday and Sunday with Bone's ability improving greatly. We were very tired upon arriving home late Sunday evening.

Tuesday 14 January We had advanced Harpoon training with a subject matter expert, George, from NWC. His training program was excellent.

Saturday 1 February The squadron had a farewell party for Bone combined with a promotion party for Killer and Decoy. It was sad to say

---

[228] Assistant Chief Operational Test Director (ACOTD) assisted the COTD in coordinating operational testing.

goodbye to Bone. He was a great part of VX-5, loved by everyone and had great potential to be an amazing fleet Intruder pilot.

Tuesday 18 February Wingnut and I prepared a PowerPoint presentation to brief the Vampire aircrew on the AX aircraft development program.

Wednesday 19 February Our AX presentation started at 8am. VX-5 aircrew discussed and debated requirements for the new aircraft until noon. This was a rare opportunity to capture the thoughts of many tactically proficient aircrew into aircraft design goals. We drafted five pivotal operational requirements for the development.

Friday 21 February Chet and I flew an Intruder as wingman for Hornet pilot Dingle on a dive-bombing mission to Superior Valley. After dropping our MK-76 blue bombs, Chet and I performed IR Maverick practice attacks.

Monday 24 February I traveled to PAX for a classified meeting with Strike ordnance engineer, Neal. He and I had a great discussion on weapon release considerations for future projects. Neal was considered a guru for the specialty and his guidance was very helpful. I went over to the Hazelrigg Hangar and saw T-45s, F-18s and a Super Tomcat aircraft.

Tuesday 25 February Wingnut and I attended an AX Action Officer meeting. Pilot Rocco and BN JB from VA-115, Duck (CVW-5 LSO), Frenchie, TOPGUN, Admiral McGinn and others were in attendance. Admiral Dunleavy provided a solid briefing on our path forward with the AX program. Five tactical aircraft contractor teams had been awarded contracts to present proposals for a next-generation, carrier-based, strike aircraft to replace the Intruder. As AX Action Officers, we would attend the briefs around the country and prepare documents that would outline the aircraft design. The Mission Need Statement (MNS) would generate an Operational Requirements Document (ORD)[229]. The ORD contained the Key Performance

---

[229] A blank sample ORD can be found here; https://www.dhs.gov/xlibrary/assets/ORD_Template.pdf. ORDs are often classified. The ORD was later replaced with the Capability Development Document (CDD).

Parameters (KPP) required of the new aircraft. Our group had direct input into the KPPs and the ORD wording.

Wednesday 26 February Wingnut and I attended more AX briefings. He presented our VX-5 perspective to the other AX Action Officers in the afternoon. Our hard work had paid off as our input was well received.

Monday 2 March The weather was quite strange for China Lake; 2000ft ceiling, 3nmi visibility with rain. Year-round clear blue skies were the norm for the high desert. Wingnut and I were scheduled for a routine training mission with a 6am brief. We decided to take advantage of the rather poor weather and executed three HI-TACAN Runway 32[230] approaches at China Lake. We entered the clouds and rain at 2000ft AGL and it was continuous up to the initial approach fix at 17,000 feet MSL.

Once back on deck, the day became quite busy. We briefed our CO on the AX developments from the prior week. A group of other folks wanted our attention on a variety of topics. Duck wanted a faxed copy of the Vampire AX briefing. We had another party to say goodbye to Bone.

Thursday 5 March Intruder test pilot and the VX-5 ACOTD, Jim and I were selected for transition to be Navy AEDOs. That was very exciting news and an inflection point in our careers.

Friday 6 March Wingnut and I flew an instrument cross country in an Intruder from China Lake to PAX via fuel stops at Dyess AFB near Abilene, TX and Dobbins AFB near Marietta, Georgia. We had a quick refuel at Dyess. We had been checking the weather at PAX throughout the journey as it was forecast to have low ceilings and poor visibility for our arrival. While getting fuel at Dobbins, we updated our weather brief for PAX with the weather office at base operations. PAX was using runway 6. The weather was forecast to be barely above minimums for our nighttime arrival with ceilings at 200 feet above the ground with a visibility of ½ of a statute mile. We were hoping to get airborne before dark but our refuel was delayed.

---

[230] The approach plate for our flightpath is here;
https://flightaware.com/resources/airport/KNID/IAP/all/pdf.

The flight to PAX was rather short and uneventful. We were excited and on top of our game as we anticipated the PAR to Runway 6[231]. The Intruder precision approach minimums for a fully NATOPS and Instrument qualified pilot and BN were 100 feet above the ground and a ¼ statute mile of visibility[232]. We had those qualifications. Fortunately, the weather was forecast to be better than those levels. Upon check-in with the PAX approach controller, he confirmed our request for a PAR noting the weather was below 200 feet and ½ of a mile visibility. We confirmed the weather situation and our ability to continue the approach to 100 feet and ¼ of a mile of visibility. We had plenty of fuel to divert to an alternate airport if we were unable to land at PAX.

The initial PAR controller provided headings and altitudes for Wingnut to fly and he followed each direction precisely. We were instructed to lower the aircraft's landing gear and flaps around 6 miles from landing and slow to our approach speed of 140 KIAS. Once stabilized at the approach speed and on course, the final controller came on the radio, "Misty 501, this is your final controller. You are 5 miles from landing, on course, approaching your descent point". Once the final controller started speaking, no verbal response was required from our aircraft provided the approach proceeded well.

We carefully listened to every word from the controller. Wingnut expertly responded to the controller's feedback on our position relative to an ideal approach. Wingnut was focused on his instruments to keep our rate of descent within +/- 50 FPM from ideal and our heading within a degree of that assigned. That was a challenging task with turbulence, rain, crosswinds and the rather archaic instruments in the Intruder. I monitored the controller's calls and Wingnut's responses while providing a second set of eyes for safety. The controller announced, "Misty 501, begin descent, on course, on glidepath". The controller continued to provide instructions then stated, "Misty 501, approaching minimums".

---

[231] An approach plate for Runway 6, NAS Patuxent River;
https://flightaware.com/resources/airport/KNHK/IAP/ILS+OR+LOC_DME+RWY+06
[232] These minimums were unique as most single-piloted naval aircraft had weather limits of 200 feet and ½ of a mile visibility per OPNAVINST 3710. I needed to monitor the approach and flight instruments, control external communications and assist in visually acquiring the runway for us to legally proceed below 200 feet.

As we approached 200 feet above the ground, I was eagerly looking for visual signs of the runway environment. There were none. As we were in a cloud, our nose-gear mounted landing light engulfed us in startling brightness. We descended through 200 feet above the ground at our nominal rate of descent while on course. We had one hundred more feet of descent taking eight seconds before we had to wave off. Descending below 150 feet above the ground, we started to break out of the clouds. It was suddenly very dark as our landing light beamed ahead of us. I visually acquired the high intensity approach lights directly in front of us at 125 feet AGL and quickly made a radio call, "Misty 501 is visual". The approach controller responded, "Copy Misty 501, proceed visually".

Wingnut transitioned to a visual approach and landed the aircraft. He cleared the runway, took off his oxygen mask and secured the flow. I could see a light, confident smile on his face as I congratulated him on the approach.

Sunday 8 March Wingnut and I organized our return flight to China Lake via Scott AFB outside St Louis and Buckley AFB in Denver. We picked up fuel at Scott. After refueling in Denver, we climbed west into the Rockies where it was snowing. Air Traffic Control (ATC) restricted our climb to 10,000 feet MSL on departure placing us in an icing layer with accumulations growing on the leading edge of our wings. That was dangerous as aircraft lose lift with wing icing and accidents can occur. We requested and were finally granted a higher altitude en route to our cruising altitude in the mid 30s. The ice on the wings dissipated at higher altitudes and while cruising at Mach 0.7. As we descended into the China Lake restricted areas, we flew around the Pinnacles to the east of the airbase before landing.

Thursday 12 March The squadron had a Hail and Farewell for Decoy, BN Bags and Jake at the Barefoot Bar. Wingnut excelled with printing a large quantity of thin cardboard masks with Decoy's face on them and distributing them all over the bar. A farewell skit honoring our departing squadron mates was performed with the crowd going wild. Great speeches were made by those departing. The night finished with a close and raucous game of CRUD. Squadron morale was on a high.

Sunday 22 March Wingnut and I reviewed our notes for the AX comparison. We put together good analysis tools that served us well when presented with new aircraft proposals and when considering the critical wording for the AX ORD.

Tuesday 24 March – Wednesday 25 March Wingnut and I attended the McDonnell Douglas AX briefs then flew to John F Kennedy (JFK) international airport, New York to attend Grumman's pitch for the AX program.

Thursday 26 March - Friday 27 March A Grumman Vice President presented an excellent opening brief. It reflected insight that came from being a lead aircraft supplier for the Navy for decades. They had supplied the A-6E Intruder, EA-6B Prowler, F-14 Tomcat, E-2C Hawkeye, and C-2 Greyhound Carrier Onboard Delivery (COD) aircraft. The briefs were polished with data to back up their proposals.

Monday 30 March I drove three hours to LA for AX briefings the next day. Wingnut and I met to discuss our latest AX thoughts.

Tuesday 31 March - Wednesday 1 April Wingnut and I attended the AX briefings at Rockwell. We saw fellow Eagles Rocco and JB again. Rocco suggested that Wingnut and I brief the Deputy Chief of Naval Operations, (Naval Warfare), OP-O7 on our AX experiences to date. Wingnut and I flew from LAX to Fort Worth, TX for General Dynamics (GD) briefs.

Thursday 2 April - Friday 3 April GD gave a good presentation on Helmet Mounted Displays (HMD).

Tuesday 7 April - Wednesday 8 April Lockheed gave their AX presentations in Atlanta. They had an F-22 in the hangar in case there was any question about their ability to build impressive aircraft.

Thursday 9 April Wingnut and I flew to LAX. I drove 3 hours south to San Diego, checked into the NAS NI BOQ, ran into Scanner and other VA-165 Boomer crews.

Friday 10 April I attended an AEDO Symposium at NI. Three-star Admiral Bowes from the Naval Air Systems Command (NAVAIRSYSCOM)[233] provided the introduction. Congressman Duke Cunningham (Vietnam War Ace, TOPGUN CO) was the guest speaker at lunch. In discussions after his speech, he commented, "AX should be supersonic but Naval Aviation is in Chapter 11" meaning bankrupt. A group of us ran into HMAC along with other Intruder crews.

Saturday 11 April An impressive crowd estimated at 5000 were present for the decommissioning ceremony of the USS Midway. The Secretary of the Navy Garrett, 15 former Midway Commanding Officers and 25 original Midway crew (plankowners) from 1945 along with representatives from Japan were present. USS Midway had been in service for 47 years changing the lives of many people including my own.

Three aircraft flybys occurred during the ceremony. The first was a three plane of WWII vintage aircraft, the second was a formation of two A-1 Skyraiders and the third was a diamond formation of four F/A-18s. As there were many Filipino friends of the USS Midway, refreshments after the ceremony included lumpia along with fruit cake and a ship's sponge cake. In the hangar bay, there were pictures from Desert Storm and picking up refugees after the eruption of Mount Pinatubo[234] near Subic Bay.

Sunday 12 April While leaving the BOQ, VX-5 squadron mate, Jake, introduced me to his brother, Dell[235] and his father, Admiral Bull[236]. Jake's father was a national hero with combat service in Intruders during

---

[233] NAVAIRSYSCOM (NAVAIR for short) is the agency responsible for purchasing, engineering and logistics of aviation products for the U.S. Navy. https://www.NAVAIR.navy.mil/

[234] The enormous eruption of Mount Pinatubo on 15 June 1991 triggered the departure of the US military in the Philippines. The eruption was the second largest of the century on earth. https://pubs.usgs.gov/fs/1997/fs113-97/

[235] Admiral Dell Bull; https://www.navy.mil/DesktopModules/ArticleCS/Print.aspx?PortalId=1&ModuleId=692&Article=2236281

[236] Admiral Lyle Bull; https://en.wikipedia.org/wiki/Lyle_F._Bull

the Vietnam war[237]. His heroism inspired the book and movie, Flight of the Intruder[238]. Jake and Dell reflected their father's patriotism and perseverance. It was an honor to meet them both.

Tuesday 14 April PK and I briefed for an Intruder flight. We carried a SLAM missile and an AWW-13 datalink pod. We took off and tested the SLAM missile as we headed southwest of China Lake. As we approached the Kern River Valley, we spotted a B-2 bomber[239] cruising along with its airborne chase aircraft. We kept our distance but watched it for a minute. PK and I conducted SLAM practice attack runs on the California City airport.

Wednesday 15 April PK and I started the day with a 7am brief for a Functional Check Flight (FCF) B profile in an A-6E. We flew to 40,000 feet in accordance with the FCF checklist and came back down at 575KIAS/M0.9 on descent. We flew by the VX-5 plaque at low level in the Panamint Valley at the precise moment another Intruder was making his pass in the opposite direction. We returned to China Lake and debriefed the flight. I quickly started a brief with Wingnut for a bombing/HARM missile flight. We took off from China Lake and transited over to Superior Valley to drop a rack of MK-76 bombs. The bomb hits were good and we started heading west to depart the bombing range and conducted our HARM missile testing before returning to base.

Thursday 16 April I provided a Desert Storm lessons learned brief to NWC staff. They shared insightful Desert Storm videos they had acquired.

Saturday 25 April I received a call from Bone. He had finished SERE school and was impressed with the training.

---

[237] Details of Admiral Bull's attack on Hanoi can be read here; https://www.usni.org/magazines/proceedings/1969/july/hanoi-tonight#:~:text=On%20the%20night%20of%2030,the%20history%20of%20air%20warfare.
[238] The Steven Coonts novel is an excellent read; https://www.coonts.com/flight-of-the-intruder.html
[239] The B-2 was beautiful to see flying! https://www.northropgrumman.com/what-we-do/air/b-2-stealth-bomber/

Wednesday 13 May PK and I attended a classified meeting in Buffalo, NY. The audience benefited greatly from PK's tactical insight. We got back to China Lake at 11pm west coast time (2am east coast time).

Saturday 16 May Twenty-five VX-5 crew and family members assembled at 11am for a road trip to Edwards AFB to watch the landing of the Space Shuttle Endeavor. The Mission Commander was Captain Dan Brandenstein[240]. We found a spot on a hill overlooking the runway. The kids ran around playing while the parents chatting in anticipation for news that the landing was imminent. As the Shuttle approached overhead, we heard the characteristic double sonic boom-boom[241]. This was probably the last Shuttle landing that most of us observed. It was great to share it as a Vampire family and a great day to be an American!

Tuesday 26 May I shared discussions with various Vampires about air-to-air and air-to-ground considerations for AX. Had a great SLAM flight with PK. We flew successful practice attacks on a California City Airport hangar, an aircraft at the field, a building at Inyokern and the 'T' on the hill outside the town of Trona, CA.

Monday 1 June We received a draft of the classified AX ORD to review and discuss. The new COTD, Intruder pilot Snow and I shared a great day flight. He was a quiet, intelligent LCDR with an engineering background.

Monday 8 June Wingnut and I spent considerable time on the AX program in preparation for a squadron murder board of our proposed KPPs.

Tuesday 9 June The Vampire aircrew had four hours of discussion regarding the AX ORD and the KPPs. There were many contentious discussions that occurred.

Friday 12 June Bert and I had a 0545 brief for a F/A-18D flight, this time carrying an IR Maverick missile and a targeting FLIR. He did a

---

[240] Captain Brandenstein is a Vietnam War decorated Intruder test pilot that served at VX-5, VA-128 and at Strike at PAX following graduation from USNTPS.
[241] Shuttle double sonic boom; https://www.youtube.com/watch?v=lNL4HHFG8H4

great job of flying the aircraft at low altitude in mountainous terrain while controlling the missile. We did practice attacks on the dam and a bridge at Lake Isabella, the rock crusher at the Owens Dry Lakebed and the Trona 'T'. After the flight, Wingnut and I briefed the CO on the latest AX aircraft developments. We had a squadron picnic and a Hail and Farewell in the evening. Craw joined us as he was going to be the best man at my upcoming marriage to Wendy in Australia[242].

Monday 15 June Wingnut and I continued working on the AX ORD. We briefed the CO on our AX ORD game-plan. I juggled final administrative details then made it to Inyokern airport for a flight to Sydney.

Friday 26 June Wendy and I arrived back in China Lake following our wedding.

Sunday 28 June I woke at 130am from jet lag, wrote journal entries and read for a couple of hours in our lounge room. Shortly before 5am, Wendy walked into the room as a 7.3 magnitude earthquake with an epicenter 150 miles south of us, started shaking the house. Neither one of us had ever been in an earthquake that large. We exited the back door, stepped out into the backyard and witnessed the flat desert floor rolling like ocean waves across a great distance. The earthquake lasted about thirty seconds with the greatest intensity being at the start.

We composed ourselves from the early morning excitement, grabbed our bags and drove out to Inyokern airport for a return flight to LAX. We were headed to Washington, D.C. where Wingnut and I were participating in the development of the new AX ORD. As Wendy and I sat in the food court at LAX at 805am, the 6.5 magnitude Big Bear earthquake hit. We stared out the large picture windows at the wings of the aircraft on the parking apron as they moved up and down in response to the quake. The aircraft engineers determined that the aircraft hadn't been harmed and flights continued. We arrived in D.C. and found our way to our hotel in Crystal City.

---

[242] I haven't included details of our courtship to retain our privacy.

Monday 29 June I woke at 4am due to jet lag. Wingnut and I met to consolidate our notes. Results from a
Cost and Operational Effectiveness Analysis (COEA) were briefed.

Tuesday 30 June Wingnut and I shared a second day of AX meetings. A division appeared in the attendees with those with fighter backgrounds insisting that the aircraft engines had afterburners to provide sufficient thrust to accelerate and turn well. The A-12 design didn't have afterburners but relied more on stealth for survivability. Some stealth proponents preferred the lower radar signature than having afterburners[243].

Monday 20 Jul I was asked to be part of the reviewing team for the ASPJ OPEVAL final report and started reading.

Thursday 23 July I woke early to complete reading the ASPJ report. A small group led by Decoy conducted a murder board in the morning.

Friday 24 July More morning discussion occurred on the ASPJ report followed by SLAM planning for an upcoming detachment to NAS Fallon on Sunday. Joe, as the OTD, planned great tests for the missile. The first set of tests were to impact targets at San Nicholas Island off NAS Point Mugu. The second tests used targets at White Sands Missile Test range, New Mexico. Our detachment at Fallon facilitated the first set of tests.

Monday 27 July There was a lot of anxiety in the ready room as we were firing multiple SLAM missiles. The missiles performed well with bullseyes but there were software anomalies that needed changes. We reported those problems back to the developers for their correction. Fixes often required months if not years to be seen in the fleet.

Monday 3 August VX-5 had a NATOPS training day. There were manpower shifts happening in the squadron and I was selected for the vacant ACOTD position.

---

[243] The push for a more agile aircraft with fighter performance eventually resulted in the A/F-X program.

Tuesday 4 August On occasion, VX-5 crew flew with NWC China Lake pilots. K9 was at NWC and we had the opportunity to run around the restricted areas in an Intruder. We hadn't flown together since March of 1988 while in VA-115. It was great to fly with him again, our last flight together.

Wednesday 5 August I visited the AX office on base to be briefed on concerns with the draft ORD. I read portions of the F/A-18 E/F Test and Evaluation Master Plan (TEMP)[244]. This document summarized all the testing that would be completed on the aircraft before delivery to the fleet.

Friday 7 August We had a Hail and Farewell for Jim, Killer and Sam at the Bowling Alley. Sam was my flight school roommate in Pensacola and more recently squadron mate at VX-5. We were going to miss him along with Killer and Jim.

Friday 14 August The ACOTD position provided an opportunity to help the other OTDs. They worked independently but training for new OTDs was lacking. The training in Norfolk didn't provide the information they needed to do their job in China Lake. I hoped to remedy that gap in knowledge.

Saturday 22 August Skyler, his spouse, Karen, Wendy and I climbed Mount Whitney. We woke up at 4am then drove for two hours to the Mount Whitney Portals. We started the climb at 7am, suffered throughout the switchbacks at 12,000 feet, reached the summit at 14,505 feet above sea level at 3pm! We left the summit around 330pm and arrived back to the Portal at 9pm. Darkness was growing as we got to the parking lot. We were pleased to be off the trail. We had dinner in Lone Pine (3727 feet of elevation) then returned to China Lake. It was a very long and crazy day!

Tuesday 25 August I worked with Bert on a flight test matrix, completed TAMPS work then got busy with the new ACOTD job.

---

[244] The TEMP was a contract between all parties regarding how an item would be tested to ensure it meets the ORD.
https://www.dote.osd.mil/Portals/97/docs/TEMPGuide/DefenseAcquisitionGuidebookCh9.pdf

Tuesday 1 September We had a Murder Board on the Final report for the EA-6B Band 2/3 jammer. Zoil and I chatted about his HARM Block IV testing. We had a farewell cake for Jim. He was headed off to a management job in Ohio. He was able to make big contributions on that program and many others during his naval career[245].

Wednesday 9 September Joe and I had our first flight together in an F/A-18D; it happened to be a night flight. It was a great change to go from being crewed together in Intruders with the Eagles to flying a Hornet with the Vampires. The aircraft was equipped with both Navigation and Targeting FLIRs. We flew to El Toro for instrument practice then returned to the restricted area to work our crew coordination with the radar and the FLIRs.

Thursday 15 October Most OTDs came straight from the fleet and had no formal flight test training. A great deal was riding on their efforts, yet they needed more information to do their job well. I wrote the OTD Training Guide and created a training course to assist them transition to the world of operational flight test.

Wednesday 28 October We had a meeting to discuss the best strategies to use when discussing the need for Operational Test (OT) with those that didn't want the time or expense of OT. I worked on NA and NVG Operational Tactics Guide (OTG)s, threat lists and simulators in the afternoon.

Thursday 29 October My detailer reported that my next duty station was NAVAIR in Washington, D.C. working engineering issues for the F/A-18 Hornet program. NAVAIR is tasked with delivering everything naval aviation needs to complete its global tasking. That is an astronomical task that I would learn to appreciate in years to come.

---

[245] Jim had a great naval career. He unfortunately passed away from cancer in 2018 at age 61; His spouse, Betty, successfully lobbied Congress for an investigation into the cause. https://ktla.com/health/ap-health/higher-cancer-rates-found-in-military-pilots-ground-crews/

Wednesday 4 November The morning dealt with SLAM and EA-6B software problems, a SA-8[246] OTG and the TAMPS ORD. I briefed at 11am for an Intruder NATOPS check flight with Deano. It was good to get out of the office. We flew a visual low level, practiced our SRTC terrain following maneuvers then went over to Superior Valley to drop six MK-76 blue bombs using visual, radar and FLIR delivery modes.

Monday 16 November I flew in an F/A-18D in the morning and drove three hours to Lemoore in the evening arriving at 10pm.

Tuesday 17 November - Wednesday 18 November I attended a Tactical Air Night Attack (TANS) Working Group (WG) meeting at Lemoore. Bullet also attended the WG.[247]

Friday 4 December The squadron had a Hail and Farewell at the Barefoot Bar for Gunner, Skyler and me. Zoil played Mr. DJ spinning music. He played John Mellencamp's 'Check it Out' and we all sang along hoping that future generations would have a better understanding[248].

Wednesday 9 December I met my predecessor at the NAVAIR Hornet Class Desk, callsign Boots. He was an ECMO with an electrical engineering background and clearly a very smart character. Our turnover started in Washington, D.C. in mid-January.

Sunday 13 December Dingle[249] and Sue hosted a party for the Vampire aircrew at their home. Wendy and I enjoyed it greatly and would miss the Vampire comradery.

---

[246] Soviet anti-aircraft missile;
http://www.armedforces.co.uk/Europeandefence/edequipment/edmis/edmis5a18.htm
[247] Vice Admiral 'Bullet' Miller's career flourished resulting in his leadership of naval aviation at the Pentagon;
https://www.navy.mil/DesktopModules/ArticleCS/Print.aspx?PortalId=1&ModuleId=692&Article=2236125
[248] John's band, 'Check It Out'; https://www.youtube.com/watch?v=8qxDBiiVjlQ
[249] Dingle and Sue loved China Lake so much that they returned later with Dingle as the VX-9 CO. He worked for Zoil who became COMOPTEVFOR and worked with Decoy who led NAVAIR.

Tuesday 15 December I started a turnover with the incoming ACOTD, Intruder pilot, Buds. I wrote a Bat Bulletin article encouraging fleet aircrew to get involved in flight test and acquisition. It provided the path to getting the needed tools to the fleet.

Wednesday 30 December Snow and I shared my last flight in an Intruder at VX-5. We flew a roundtrip to NAS Fallon and checked out at the snow around Mammoth on our return.

Thursday 31 December A group of us headed to Mammoth to go skiing. We needed my 4WD to get to the Vampire lodge as the area had recently received a large dump of snow. We enjoyed dinner at a Mexican restaurant and champagne at midnight to celebrate the New Year. We had a great weekend with friends skiing fresh snow.

In the week that followed, we were able to finalize administrative details, pack up our belongings and share special times with close friends at China Lake.

# Chapter 13

## F/A-18 Hornet Engineering, NAVAIR, Washington D.C.

*We received a naval message that a Hornet crew in an F/A-18D had inadvertently ejected from their aircraft on a shore-based landing rollout. With over a thousand Hornet aircraft built and over a million flight hours, this had never happened.*

Monday 11 January 1993 Wendy and I left China Lake, CA destined for Washington, D.C. It was a long journey but provided beautiful and interesting views across the vast USA.

My first assignment as an AEDO was working as a Systems Engineer for the F/A-18 Program Office inside NAVAIR. I inherited the following projects from Boots; the APG-73 Radar Upgrade (RUG) Program[250], an improved bird-strike resistant windscreen[251], Cockpit Video Recording System (CVRS), the On-Board Oxygen Generation System (OBOGS)[252], the OBOGS Solid State Oxygen Monitor (SSOM) upgrade, and various crew system interface issues. Other projects were managed as they arose including a couple of highly classified ones.

Before the government bought something, a need was identified. With large expenditures, the government formalized the requirement with a MNS and ORD. Once an ORD was established, the government generated a specification that listed specific requirements for the identified need. Once a specification was written, the Statement of Work (SOW) and Request for Proposal (RFP) were written. Companies

---

[250] The RUG Phase I program development and procurement cost was $1.4 Billion (FY 95 $) making it an ACAT I, Major Defense Acquisition Program. The program purchased hardware for 270 new and existing F/A-18C/D aircraft. The E/F aircraft were also equipped with APG-73 radars.
https://media.defense.gov/1994/Dec/30/2001714856/-1/-1/1/95-070.pdf
[251] A Hornet pilot was killed when a bird came through his windscreen while he was flying at low altitude at 465 KIAS. This increased the bird-strike resistance requirement leading to a new windscreen design.
[252] The OBOGS system took jet engine bleed air, put it through scrubbers and extracted oxygen that was passed to the aircrew for breathing through their masks. Older aircraft used bottled oxygen systems for the aircrew.

responded to the RFP with their proposals to fulfill the need. The proposals needed to satisfy the conditions of the SOW including the delivery schedule and meet the product specification. This was the long-standing process for government purchases. Things could be streamlined depending on the time criticality, the specifications and the cost involved. There was a push to purchase Commercial Off The Shelf (COTS) products as well as abandon all government specifications as they were deemed too burdensome for the vendors.

Work at NAVAIR was exciting yet challenging with long hours. Purchasing and maintaining equipment for naval aviation is not an easy task. NAVAIR is a giant juggling exercise that balances engineering, logistic and maintainability requirements, testing needs, schedule demands, funding priorities, various acquisition legal limitations and schedules to cite just a few. At times, everyone's patience was thin in a tense atmosphere. Contracts legally required competitive acquisition which took time and great effort. There was a steady flow of documents and emails to read and write. The documents included product specifications, SOWs, RFPs, contractor responses to RFPs, contracts, progress reports from contractors and government agencies, technical trouble reports, fleet mishap reports and message traffic identifying technical problems, as well as Preliminary Design Review (PDR) and Critical Design Review (CDR) slides and minutes.

NAVAIR work demanded an extensive travel schedule. It was not unusual to be away from Washington, D.C. for 2 weeks out of every month with frequent trips to China Lake. It is important to note that we didn't have the luxury of Skype, Zoom or other easily accessible, effective, long distance group communication tools. Great amounts of taxpayer money were dependent on doing business well and that required travel to ensure contracts were properly executed. We traveled across the country to look one another in the eye and to agree on the path forward. We needed to ensure that the promises we made to one another were clear, unambiguous and the required fleet products were delivered on time. A sample of the work and climate at NAVAIR is provided in the details that follow.

I routinely woke at 5am, ran for 30 minutes, left home at 630am and arrived at my desk at 7am. It was common to leave work at 630pm to arrive home at 7pm. There were nights I didn't get home until 9pm or later.

Thursday 21 January My first day at NAVAIR was jam-packed with activity. The morning required administrative chores for security and meeting various staff members. I was escorted to the F/A-18 Class Desk area on the 11th floor of the Crystal City office building. The floor was mostly an open-plan design with various aircraft engineering departments working in adjacent cubicles. Smiley was the A-6E / EA-6B Class Desk Officer[253] and sat about 20 feet from my desk. It was great to see his perpetually, welcoming smile.

I met my boss, the F/A-18 Class Desk Officer, Commander Gib Godwin[254]. He was a Desert Storm combat pilot and served as the CO of the VFA-192 Golden Dragons aboard the USS Midway and Independence. Gib worked for Captain Steidle who served as the Hornet Program Manager (PMA). The PMA managed all aspects of the F/A-18 including supervising the Assistant Program Manager of Logistics (APML). As the Hornet was flown by many other countries, the PMA was responsible for ensuring the global needs of the program were met. At this time, the Hornet was flown by the US Navy, Australia, Canada, Spain and Kuwait[255].

The afternoon was spent with Boots. He initially enlisted in the Navy and was initially an electronics technician aboard nuclear submarines. Boots' success resulted in his selection for officer promotion and he obtained a Bachelor's degree in Electrical Engineering. Boots went to flight school and trained in the EA-6B electronic warfare aircraft following his NFO winging. We discussed our projects for hours. There were so many details, concerns, problems, plans, structural, electronic, testing or installation issues. The job got easier with time but was quite confusing at the beginning.

---

[253] The Class Desk Officer was responsible for all engineering issues related to a particular aircraft design. They led staff that were distributed across the country that supported the fleet aircraft.
[254] Admiral Godwin had a very successful career with the Navy; https://wikimili.com/en/James_Godwin
[255] Finland, Malaysia and Switzerland purchased Hornets in years to come.

Friday 22 January Boots and I reviewed information in preparation for a Systems Safety Working Group (SSWG) meeting. We discussed the urgency of getting video cameras and new tape recorders into all F/A-18 aircraft as soon as possible. Desert Storm had driven home the requirement for all combat aircraft to have video footage of weapon impacts. Due to the nature of the display in the Hornet, we couldn't record a digital signal but needed to record the video images from the cockpit displays with a camera in the cockpit. The CVRS project planned to purchase hardware and have it installed in fleet aircraft.

Hardware that flew in an F/A-18 had to endure a demanding environment of vibration, acceleration, shock and temperature extremes. Any item inserted in the cockpit could not pose a risk to the aircrew nor compromise a successful pilot emergency ejection. The cameras were to be placed on the canopy rails behind the pilot's shoulders. They couldn't obstruct his view outside the cockpit nor intrude as he was flying and fighting the aircraft in the most demanding of conditions. The cameras needed to perform well in bright daylight as well as in a dark cockpit at night. The camera and recorder installation designs were completed at Navy facilities at Warminster, PA and NAS NI in San Diego.

The cameras and recorders were to be competitively procured by NAVAIR using COTS items to streamline the acquisition timeline and cost. Industry claimed they had amazing technology sitting on a shelf for the DOD, but our acquisition process was too laborious to deliver the items to the warfighter. We streamlined our acquisition processes to acquire these COTS items when feasible. CVRS was an attempt to expedite hardware into the cockpit in record time at a reduced cost. NAVAIR managed the camera and recorder procurement. I was responsible for coordinating the engineering inputs.

An Engineering Change Proposal (ECP) to approve the CVRS installation was routed through various departments inside NAVAIR for agreement and sign-off. An approved ECP was required to install the hardware in the aircraft. The ECP approval process could take on a life of its own and take months or years if not actively managed. Boots took me around to meet the NAVAIR staff that would approve the ECP. Despite being on the 11th floor and other staff being on various floors of the building, Boots always took the stairs instead of the elevators. He was a long-distance runner, had a wiry frame and managed the stairs without skipping a beat. He kept talking program details as we climbed the stairs.

Monday 25 January The day was filled with more information from Boots. We prepared for our attendance at the SSWG in Jacksonville, FL the next day. We left work at 2pm, flew from Washington National airport to Jacksonville then rented a car and drove to the NAS BOQ.

Tuesday 26 January The SSWG had participants from all over the country. Topics that caught my attention were engine fires, deteriorating Flexible Confined Detonating Cord (FCDC), brake problems, windscreen concerns and others. The FCDC was a fully qualified yet aging explosive cord that facilitated jettison of the canopy in the event of aircrew ejection. The life limit of the FCDC had been extended but the thermal, vibration and flexing environment of the cockpit was generating internal cracks compromising integrity. A replacement was needed before an accident happened. Fortunately, a Thin Line Explosive (TLX) cord appeared to be a qualified replacement.

Wednesday 3 February We had a flight clearance meeting for CVRS at 10am. Crew environmental system discussions also took place. There were engineers that knew their profession well and gave quick, accurate response to questions. They were great to have as allies.

Bone called with the great news that he was going to join my old squadron, the VA-115 Eagles aboard the USS Independence in Japan.

I attended an interesting RUG reconnaissance meeting. Afterwards we had a meeting with tactical aircraft windscreen design experts from Wright Patterson AFB. They assisted in defining a bird-strike resistant windscreen design specification. I picked up orders and plane tickets for a flight to McDonnell Douglas Aircraft Company (MCAIR), St Louis, MO.

Friday 12 February MCAIR Hornet Crew Systems Team Leader, Lyle[256] and I met to discuss various issues. He had prepared a folder laying out in appropriate detail all the current issues requiring my attention with his recommended solutions. Lyle knew that I needed to make technical decisions and provided just enough history and information to make the decisions easy. He proved to be invaluable through my time at NAVAIR and in the years to come. Lyle introduced me to the other staff members supporting his Crew Systems group. They were a great team and I appreciated their efforts in working a variety of challenging issues. We visited their static cockpit and ejection seat displays that could help resolve design issues.

Monday 15 February I flew to LAX to attend a Hornet radar upgrade meeting. There were reports that China Lake was having development problems with the APG-73 RUG program. Gib asked me to sort out the details.

Tuesday 16 – Wednesday 17 February We had Technical Coordination Meetings (TCM) with MCAIR, radar manufacturer, Hughes and government representatives to discuss the RUG progress. The group had mixed opinions on the status. I drove to China Lake in the afternoon after the meeting. It was great to be back in the wide-open spaces of the high desert of California and to catch up with my Vampire mates.

Thursday 18 February I met with the RUG government leaders and engineers onboard NAS China Lake. Good discussions with various government employees were held. The progress of the program remained unclear as I received conflicting information. Developmental and initial operational testing was scheduled for completion by October 1995 to support a crucial production decision. This date appeared impossible due to developmental delays, but managers fiercely defended the schedule[257].

Friday 19 February I caught a 720am Northwest flight return to Washington, D.C. via Memphis. One of the challenges with working across the USA was the distance involved. I used Washington National

---

[256] Lyle possesses great knowledge, experience, wisdom and insight into tactical aviation. He knew how to get things done well and quickly.
[257] The production decision eventually shifted to July 1996 due to various development problems.

airport as it was convenient from NAVAIR and my home. There were no non-stop flights from the west coast to National; an intermediate stop was required. An early morning flight from the west coast arrived at National around dinner time.

Monday 22 February Boots and I continued our turnover with the various active issues. Boots, PMA staffer, Bill and I traveled to Warminster, PA (2.5-hour drive north) in the early evening. We discussed work priorities throughout the journey.

Tuesday 23 February We visited the Naval Air Warfare Center (NAWC) Warminster to discuss the CVRS camera installation with large group of engineers and designers.

Thursday 25 February We had a big field activity meeting at NAVAIR with Gib laying down the ground rules about who works for who and how he needs priorities to be set. The field activities were those government agencies located around the country that were funded by NAVAIR. There were many managers who didn't want to hear Gib's message as it compromised their autonomy, despite answering to him for their funding.

Thursday – Friday March 4-5 Smiley and I attended a Systems Engineering Course for new arrivals to NAVAIR.

Monday – Wednesday 8-10 March The Systems Engineering Course continued. The textbook was quite good[258]. It was a great course and gave everyone insight into the operation of NAVAIR, how things get designed, manufactured and deployed and how to manage the process from cradle to grave.

Friday 12 March Today was Boots' last day at NAVAIR. We had more discussions on the windscreen plan. It had been a wild turnover with Boots, but I needed to sort out the future of our projects.

---

[258] J. Lacy, *Systems Engineering Management, Achieving Total Quality*, (New York, NY, McGraw Hill, 1992)

Monday 15 March My first day on the job without Boots' guiding hand went well. I was tasked with throwing a farewell party for Boots and collecting money from the staff.

Tuesday 16 March I covered various topics of interest at work then set up the party for Boots. He had important work to do at the Defense Plant Representative Office (DPRO) in St Louis.

Friday 19 March We had another budget meeting then I briefed Captain Steidle on OBOGS problems and solutions. He wanted to correct the problem ASAP.

Monday 22 March We had another meeting with the PMA on OBOGS then had a CVRS camera procurement meeting. The windscreen RFP was put out for public response. I discovered a discrepancy with the RUG development plan leading to many discussions over the weeks and months ahead.

Tuesday 23 March Gib and PMA Reps Kathy, Mike and I had a meeting to discuss the RUG program. Another meeting was scheduled for Thursday with the PMA.

Friday 26 March We had a RUG Phase III meeting for the Advanced Electronically Scanned Array (AESA) radar hardware. The AESA radar would correct numerous issues with the mechanically scanned APG-65 and APG-73 RUG but technical concerns made it a riskier program.

Monday 29 March The RUG program was heating up. Gib wanted a meeting of significant players on Thursday in Washington, D.C. He wanted to bring representatives from Hughes Corporation in L.A., government managers from China Lake, aircraft integrators from McDonnell Douglas in St Louis, flight test personnel from PAX and NAVAIR staff together to confirm the program status. I left work at 5pm and flew from National Airport to Davenport, Iowa for an OBOGS design review.

Tuesday 30 March Lyle and I attended a meeting at Litton, the manufacturer of the OBOGS oxygen sensor. The sensor warned the pilot if the oxygen level delivered to the cockpit was low. Litton was catching heat from the Navy and Marine Corps for problems with the OBOGS system and didn't appreciate being the center of attention. Lyle and I had already discussed the best course of action to resolve the fleet problems.

The original OBOGS sensor was a state-of-the-art piece of hardware when first introduced to the fleet. Unfortunately, it was a maintenance-intensive piece of equipment. If it wasn't maintained and supported properly, it became unreliable risking aircrew's lives. The new SSOM was designed to be very reliable and eliminate the need for intermediate level maintenance. We needed a plan to quickly get the SSOM into new production aircraft at MCAIR and retrofitted to Hornet aircraft deployed worldwide.

In the meeting, we proposed that production of the SSOM be accelerated, eliminating intermediate maintenance and expediating delivery to the fleet within a year. Litton representatives were stunned. They were contracted by NAVAIR to follow a slower path. I asked if they could expedite the SSOM replacement and deliver units quicker if we renegotiated the contract. They confirmed that they could. Doing so would turn around their tarnished reputation in the fleet. We all agreed to the revised schedule.

The Litton representatives were so pleased they asked the company president to meet Lyle and me. The Litton President listened to our progress and agreed to use $500,000 of corporate funds to expedite the SSOM to the fleet. He requested that the NAVAIR contracting officer[259] modify the contracts to bring the money forward rather than stretched out into the future. I agreed to the changes as the PMA wanted the problem fixed ASAP. We were impressed that Litton made the changes in good faith. I took an action item to organize the contract modification when back at NAVAIR. That had been a great meeting with credit to Lyle and Boots who laid the foundation for the success.

Wednesday 31 March I sent out an OBOGS email to update everyone on the progress made in Iowa. We then had an intense meeting with a group of folks about the RUG radar program. There were

---

[259] Contracting officers are those individuals that sign contracts on behalf of the United States Government.

misunderstandings occurring and poor communication amongst the team. The Thursday RUG meeting was delayed.

Thursday 1 April The PMA requested an update on the RUG program. He shared my concerns and requested a technical status update by Tuesday. That required a significant effort to complete. I discussed the RUG technical issues with the NAVAIR experts then read the wording in the RUG contract.

Tuesday 6 April RUG principal players from all over the country met in a large, executive boardroom with the PMA, other senior naval officers, senior RUG radar engineers and managers. Cathy welcomed everyone to the meeting and handed it over to me to discuss the technical challenges. I outlined the problems and expressed the urgency in understanding the risk of successful project completion along with the schedule. The program ORD identified performance required for developmental and operational testing. The projected RUG maturity wouldn't meet the planned testing timeline so something had to give.

The F/A-18 software development manager from China Lake, Rich, presented a revised plan for further development. It wasn't what we expected but at least there was a clear understanding among the team members about the path forward. Many more meetings over the next couple years kept everyone apprised of the development of this critical program.

Monday 12 April The Class Desk staff had a meeting with the CVRS team. I had later discussions with crew systems experts on NVGs in F/A-18 A/B aircraft with unmodified lighting and glare concerns. VX-5 were willing to do an operational assessment for us.

Wednesday 14 April I visited Hornet squadrons at Beaufort speaking with maintainers and aircrew about OBOGS problems. They were struggling with OBOGS due to poor hardware performance, publications and training. I helped with strategies for maintenance then explained our plan for the expedited SSOM replacement.

Friday 16 April The Australian Hornet engineering team and I met to discuss the bird-strike resistant windscreen. They were very interested in upgrading their windscreens after losing an F-111 crew[260] and aircraft following a bird-strike on a low level in Australia. The mishap was a tragic loss for their country.

Wednesday 21 April We had a meeting with the China Lake RUG team and Hughes contractor near LAX. The technical requirements with the updated delivery schedule remained a challenge. Fortunately, everyone was seeking a path forward to success.

Monday 17 May The Wright Patterson AFB windscreen team and I reviewed the contract for the upgraded windscreen vendor. The new design featured a laminated polycarbonate material with a titanium frame to act as a shock absorber.

Tuesday 8 June The first bird-strike test of the new windscreen successfully occurred at 290KIAS at Arnold AFB, Tullahoma, TN. Higher speeds were scheduled to follow.

Thursday 10 June I traveled to Arnold AFB in Tennessee to witness a second bird-strike test of the new windscreen design. Arnold served as the testing laboratory for many defense-related projects as they had a top-notch staff and superb hardware.

The new windscreen was located on a Hornet forward-fuselage static fixture at the end of a large building. The fixture had a canopy and bottom rail to realistically support the windscreen. It was important to test the new windscreen with a configuration identical to the aircraft. The windscreen reacted with the installation structure as it flexed under the load of the bird impact. The birds used for testing were dead, thawed, 4lb chickens. The chicken needed to be thawed and at room temperature to ensure it did not possess the penetrating projectile qualities of a frozen chicken. Surrounding the windscreen were stands with very bright lights and high-speed film cameras. The lights and cameras came on a few seconds prior to the bird being shot at the windscreen. The cameras consumed large amounts of film for each second of time they were running to capture impact details. If the lights and cameras didn't turn on

---

[260] Shorty was the pilot in the F-111 accident. He was in the class ahead of me at USNTPS and a great guy.

before the shot, the shot countdown was stopped and recommenced after the problem was resolved.

Centered six feet in front of the windscreen was a metal barrel with a six-inch inner diameter. The barrel served as the pointing mechanism that aimed the chicken at the windscreen. The barrel stretched one hundred feet away from the windscreen back into the test laboratory. The chickens were propelled down the barrel using compressed air like an air rifle. Large storage tanks with quick opening valves provided the compressed air. The chicken placement in the barrel determined the exit velocity from the barrel. The longer the distance inside the barrel, the faster the chicken departed the barrel. There were many preparations before a chicken shoot and all the instrumentation in the lab had to be working flawlessly. When all things looked good, a countdown was commenced from one minute prior to the chicken firing. A few seconds before firing, the lights around the windscreen came on and the high-speed cameras started capturing the event.

The firing of the chicken came off as a loud bang. After the shot and once the test site was declared safe by the lab staff, engineers were allowed into the area of the windscreen to inspect the damage. After a test, there were bird pieces everywhere including on the plastic sheets behind the windscreen along with the smell of dead chicken. Neither were enjoyable experiences but necessary. The speed for the test was 340 KIAS. The windscreen didn't look good after the test. More analysis of the film was required. I spent the night in the Arnold AFB BOQ.

Friday 11 June It was determined that the windscreen test was a failure at 340 KIAS and the windscreen was assessed as having a 320 KIAS capability. That was unacceptable as the specification for resisting the impact of a 4lb bird was 475KIAS. A meeting was held with the windscreen manufacturer. I expressed the government's disappointment with the test result, encouraged them to quickly determine the cause of the failure and remedy it. They understood and promised to deliver to the specification in short order.

Thursday 24 June We finally installed an interim camera on a Hornet at PAX to validate the specification requirements for the CVRS camera purchase. My contractor support, Joe, helped pull it together. He was a retired navy aviation maintenance man and knew how to get things done.

Monday 28 June We juggled topics of a new TLX line, RUG progress, and the Deployable Flight Incident Recorder Set (DFIRS). The DFIRS design jettisoned a flight data recorder out of the Hornet in the event of an aircraft accident identifying the location and details of the crash. This was the military equivalent of the civilian airliner 'black box'.

Tuesday 29 June I drove over to the Office of Naval Intelligence (ONI) to discuss classified, technical aspects of air-to-air radars. An ONI analyst, Mark, was very helpful in highlighting various aspects of the Hornet air-to-air radar.

Tuesday 6 July I communicated classified discussion details from the ONI analysts to various interested parties prior to another China Lake RUG meeting.

Monday 12 July I caught a morning flight to St Louis. We had a meeting to discuss the Advanced Tactical Airborne Reconnaissance System (ATARS) and ejection seat meetings followed. I assessed a helmet-mounted tactical display in a simulator. It had a lot of potential.

Wednesday 14 July I started work on a new project, the Positive Identification System (PIDS). PIDS provided the Hornet with an ability to determine if other aircraft were friend or foe. The program was to satisfy a Congressional mandate to reduce losses between friendly forces (referred to as blue-on-blue losses).

Thursday 15 July RUG engineers met with representatives from the John Hopkins Applied Physics Lab (JHAPL) to discuss radar technology.

Wednesday 21 July - Thursday 22 July We started a morning classified RUG meeting in China Lake. Mark from ONI, a U.S. Air Force radar expert and representatives from TOPGUN were in attendance. Excellent briefs on radar design and performance were given. Top Gun gave an excellent tactical brief. It was a challenging meeting as the content was highly technical and several attendees didn't want to accept what they were hearing. Disagreement occurred as the attendees tried to establish priorities for future development.

Thursday 5 August A four-day acquisition course concluded in the morning. There were attendees that verbally attacked the operational flight test community. I listened intently on their comments but couldn't resist responding. A few of the critics had open and receptive minds and gained a new perspective.

Tuesday 31 August A F/A-18C/D Program Management Review (PMR) occurred at MCAIR, St Louis. A representative from China Lake spoke about the wonderful Hornet targeting FLIR pod. After listening to his presentation, I spoke up noting that the resolution of the FLIR was insufficient for target acquisition and attack in many combat scenarios. Additionally, there weren't enough FLIR targeting pods in the fleet causing training and logistics problems. Discussions regarding a replacement for the FLIR pod started, culminating in the Advanced Targeting FLIR (ATFLIR) that is used throughout the fleet today[261].

The Defense Science Board released its findings regarding the development and acquisition of the A/F-X aircraft. Despite supporting the need for the aircraft, the F/A-18E/F was deemed sufficient for the near-term needs of naval aviation. There were insufficient funds to pursue the A/F-X concept and it was cancelled[262]. This was heartbreaking to hear for crews all over the country after all the effort expended to develop a replacement for the Intruder and A-12. I continued to see Wingnut at China Lake when on travel. He enjoyed a great career with the Navy, then surpassed it working as a Boeing test pilot[263].

Thursday 23 September We had a 2pm meeting to discuss the ATARS data link interface with the Joint Services Imagery Processing

---

[261] Deployed F/A-18E/F crews currently enjoy the advantages of the ATFLIR (Elward, 2012, pp. 168-169).
[262] The A/F-X cancellation is discussed here; https://www.globalsecurity.org/military/systems/aircraft/a-x.htm
[263] Wingnut instructing in the Boeing 787; https://www.youtube.com/watch?v=-s9ynMnPdCQ&t=22s

System (JSIPS)[264] to allow real time reconnaissance data to be passed to ground stations or Navy ships.

Wednesday 13 October I attended a digital map meeting at 830am followed by a RUG teleconference with China Lake at 930am. I met with an analyst from ONI at 11am, then with NAVAIR reliability engineer, Steve[265], at noon to discuss the RUG TEMP.

Thursday 4 November We had a meeting with PAX engineer, Shahram[266], who was tracking aircraft faults using Built In Test Logic Inspection (BLIN) codes. The work was inspirational and a fresh look at a method to improve fleet aircraft reliability and availability.

Tuesday 16 November I attended an AEDO Symposium. Many high-ranking Navy officers including the CNO were in attendance. My VA-115 and TPS CO, Dusty, was present. He mentioned that Vicki was not well and in line for a heart transplant.[267]

Tuesday 23 November I returned to Arnold AFB at Tullahoma and witnessed a successful bird-strike test. After the previous test failure, it was important to view the second test to verify that the contractor had corrected the deficiencies in the new design. The change involved a slightly modified windscreen arch to provide better flexure during bird impact.

Wednesday 1 December We had a morning RUG teleconference with China Lake. I called the DPRO at MCAIR in St Louis to discuss bird-strike resistant windscreen fit problems on new production Hornet aircraft. The windscreens were being delivered to MCAIR as Government Furnished Equipment (GFE) for incorporation into new

---

[264] JSIPS allowed processing of Hornet ATARS intelligence imagery. https://fas.org/irp/program/collect/atars.htm
[265] Steve is a close friend who enjoyed a great career at NAVAIR managing aerodynamics projects and teams with a passionate commitment to the fleet.
[266] Shahram enjoyed great success in his career with NAVAIR.
[267] Vicki received her heart transplant after many years of waiting and her joy of life continues with Dusty. Vicki and Dusty captured their amazing heart transplant journey in their book, Heartspeak. It is an inspiring read that adds meaning to life; https://www.amazon.com/Heartspeak-Vicki-Rhoades/dp/1413425941#customerReviews.

aircraft. Boots was working the issue. He was a great help in ensuring the windscreen program stayed on track.

Friday 28 January I flew to St. Louis for meetings with Lyle. We went over to the lighting mockup to evaluate NVGs in pre-Lot 12 cockpits.

Thursday 17 February Captain Dyer, the new Hornet PMA, held an all-hands meeting and provided an inspiring and motivational speech.

Sunday 13 March I joined the Andrews AFB Aero Club. While working at NAVAIR, I missed flying. The Andrews Flying Club was an excellent flight school. The instructors were great and flying from a military airport in Class B airspace was valuable training[268].

Monday 14 March I attended a Hornet PMR in St Louis. It was a good meeting with Gib and our Deputy Class Desk Officer, DJ[269] in attendance. CVRS and the new windscreen were seen as successes.

Monday 11 April When things went wrong in the Hornet fleet, they contacted NAVAIR and MCAIR for help. At times, the problems were simple to resolve and a remedy could be put in place quickly. Other times, the solution required weeks, months or possibly years to correct. Whenever the problem occurred inside the aircraft cockpit, it took longer to remedy. There were many human factor considerations to be made and many individuals had to be part of the conversations.

We received a naval message that a Hornet crew in an F/A-18D had inadvertently ejected from their aircraft on a shore-based landing rollout. With over a thousand Hornet aircraft built and over a million flight hours, this had never happened. We needed to determine how it occurred, quantify the probability of a reoccurrence while informing the fleet of the risk. As he was the MCAIR manager for Hornet crew

---

[268] I completed my Private, Instrument and Commercial Pilots Licenses at Andrews AFB from 1994-1996.

[269] DJ was a F-14 RIO, cross-trained to be a F-14 pilot, attended Naval Post Graduate School then USNTPS. He was a pleasure to work for and eventually led NAVAIR and the F-35 program. https://www.history.navy.mil/research/library/research-guides/modern-biographical-files-ndl/modern-bios-v/venlet-david-j.html

systems, Lyle stepped up to the task. Gib gave Lyle personal directions to do whatever it took to solve the problem immediately.

It was common in tactical aircraft for pilots to pull the control stick back into their laps on shore-based landing rollouts to use the horizontal stabilizers at the rear of the aircraft as air brakes. Once the aircraft had slowed down and the stabilizers were no longer helping to reduce the speed, the stick was returned to the neutral position. The pilot of the F/A-18D crew followed this procedure and when the stick was returned to neutral, the crew were ejected from the aircraft descending back to earth via parachute. They were not injured in the ejection, the aircraft suffered minor damage but we clearly had a problem that needed urgent attention.

The ejection seat was designed and manufactured by the Martin Baker (MB) corporation of England as procured and approved by NAVAIR. The seats were provided to MCAIR for aircraft production as GFE. As such, Lyle and his team had no authority over the design of the seat or the firing handle but worked very closely with both NAVAIR and MB to address the problem. While cost and logistics issues were certainly a part of the analysis, aircrew safety was the highest priority. An effective solution was sought that could be implemented very rapidly to help protect the fleet aircrews and mitigate any possibility of an aircraft loss due to the problem. The events that follow occurred over months and it took more months to manufacture and deliver new handles to the entire fleet.

It was not immediately clear what caused the inadvertent ejection. Lyle and his team initiated an intense evaluation of how this could have occurred. They noted that the rubbery ejection seat handle could grab the plastic, ridged A/A Weapon Select Switch on the control stick when the ejection seat was raised near the upper limits of travel. The seat could be raised or lowered with a switch that controlled the electric seat motor. The interference between the A/A switch and the ejection seat handle was a significant problem that needed to be eliminated. The height variation of the seat couldn't be changed as it was essential to accommodate pilots with diverse anthropometric measurements. Changing the A/A Weapon Select Switch was not desirable as it was carefully engineered to be effectively used in combat with pilots wearing gloves. The ejection seat handle had a long, narrow, oval shape. It was longer than most ejection seat handles in other tactical aircraft. Once the handle-switch interference was identified, it seemed unusual that it hadn't been previously corrected.

As with all incidents of this nature, a formal naval investigation took place. The conclusions of the naval investigation matched the MCAIR conclusions. The Hornet fleet was quickly notified of the issue and how to avoid it.

Lyle's team brainstormed solutions to remedy the problem. In developing possible solutions, the team needed to consider the urgency to protect the aircrew while considering the cost and logistics of correcting the problem for aircraft located all over the world. This included those aircraft on deployment aboard an aircraft carrier in the Middle East and foreign customers. The ejection seat handle needed to be shorter to eliminate any possibility of handle-switch contact in the future.

The possible solutions included adding hardware to the existing handle to eliminate the interference. This solution was attractive from a logistics standpoint as the hardware could be quickly manufactured and transported to the squadrons for incorporation onto the seats. Another solution was a new handle with a shorter, wider profile and shape. After considerable effort with design options, we decided to defer the design decision until after a fleet consultation. I took prototypes of various handle designs to MCAS Beaufort and NAS Jacksonville and asked Navy and Marine Hornet pilots for feedback. The shorter and wider MB handle was the more elegant and preferred solution.

The original and replacement Hornet ejection seat handles.

Tuesday 28 June I flew from National to New York LaGuardia and rented a car to attend a CVRS camera PDR on Long Island. A successful CDR was conducted a short time later with the camera approved for production.

Friday 9 September I attended a PMA-called meeting to examine new targeting FLIR technology. Seven different avionics vendors gave us briefings on their proposed replacement for the existing Hornet targeting FLIR. There were excellent briefs and the advertised resolution was far superior to the Intruder FLIR.

Saturday 15 October I was studying my instrument flight theory work when an email arrived regarding a VA-115 aircraft mishap. I opened the email, began reading and was overcome by grief. Bone was flying on an approved low level in Japan with his BN, Lieutenant John 'Chowda' Dunn on 14 October. He tried to maneuver out of a box canyon and didn't safely complete the reversal. The aircraft crashed into a lake killing both men. I was devastated by the loss of my close friend. Bone and I shared flying in A-4s and Cessnas, camping, snow and water skiing and just hanging out together. He was an awesome, always-positive guy that would be sorely missed by a great number of friends, family and squadron mates[270].

Monday 24 October A vendor was selected for the CVRS recorder. After several weeks, several government personnel attended a PDR at their factory in California. Later, we had a CDR and the design was approved for production.

We had a big meeting with various interested parties from around the country on how we would install the CVRS cameras and recorders into the Hornet. It is one thing for NAVAIR to purchase the recorders and cameras, but we needed the Naval Aviation Depot (NADEP) to manufacture brackets, wiring harnesses and installation kits for the depot level installation teams. NADEP produced kits and plans and teams

---

[270] Bone and Chowda's names will be engraved on the Intruder Tributes at NAS Whidbey Island, at the Pacific Coast Air Museum northwest of San Francisco and adjacent to the USS Midway Museum in San Diego. Their deaths remain a great tragedy for those who knew them.

started installing the system into the fleet aircraft. At the end of the project, NAVAIR saved five million dollars by procuring GFE equipment (cameras, recorders and install kits) and installing them with government labor instead of contracting the job out.

The year of 1995 was spent working various programs and getting the following modifications into fleet and new production aircraft; CVRS, the bird-strike resistant windscreen, the OBOGS SSOM, the new ejection seat handle, PIDS, Combined Interrogator Transponder (CIT) system, Multifunctional Information Distribution System (MIDS)[271] and the Embedded Global Positioning System (GPS)[272] INS (EGI).

I left F/A-18 engineering at NAVAIR hopeful our projects would enhance the warfighting capability of the fleet[273]. The APG-73 RUG radar serves the fleet well despite the delays and challenges in early development. The revised AESA radar antenna addressed various concerns raised by TOPGUN and ONI and added great mission flexibility. The ATFLIR is an invaluable targeting sensor for fleet crews. PIDS was installed in fleet aircraft and continues to reduce blue-on-blue losses today. MIDS provides situational awareness for crews by providing a battle scene overview of friendly and enemy forces. The CVRS system allowed cockpit video to be recorded as designed despite growing pains. The bird-strike resistant windscreen proved to be a success and was installed on fleet F/A-18 aircraft around the world. The OBOGS SSOM fixed a potentially lethal fleet problem and we were able to implement the change in record time. The ejection seat handles were promptly exchanged eliminating the interference with the control stick. The TLX line was installed in the canopy eliminating the old detonator cord. Classified projects were designed, tested and implemented into the fleet. It hadn't been an easy tour, but productive changes had been made.

---

[271] MIDS provided a birds-eye view of a battle area in the tactical aircraft cockpit using data provided by outside sources like ships or other aircraft like the E-2C Hawkeye.
[272] The Hornet historically had an accurate Ring LASER Gyro INS but GPS was needed for upcoming targeting.
[273] Several of the projects;
https://www.globalsecurity.org/military/library/budget/fy1997/dot-e/navy/97fa18cd.html

# Chapter 14

## Super Hornet Experimental Flight Test 1996

*The solution took years and countless hours of computer simulation, wind tunnel analysis and flight time to define and implement. Critics of the SH program in Washington, D.C. tried to use the problem to kill funding for the aircraft. Fortunately, the aircraft, the ITT and MCAIR/Boeing survived the challenge.*

The first seven Super Hornet flight test aircraft in front of the Hazelrigg Hangar at NAS Patuxent River. Photo by Kevin Flynn.

In late 1995, as my three-year tour at NAVAIR was nearing completion, I heard that the F/A-18E/F Super Hornet (SH) flight test team needed a Government Flight Representative (GFR).

A GFR provides onsite oversight of a contractor's flight program to ensure that all safety and contractual regulations are adhered to resulting in reduced risk and enhanced safety. There are many requirements that need to be met to conduct commercial flight operations safely and efficiently. It is easy to forget requirements as pressure grows to fly. The GFR reviews the contractor's safety programs along with preparation for each flight. The preparation includes an approved plan and flight clearance for the planned maneuvers; qualified, flight current and properly briefed crews and test team along with an aircraft with

254

complying maintenance documentation. Once the GFR is satisfied that all requirements were met for a safe flight, a form is signed with the government assuming liability for the loss of the aircraft should a mishap occur. Essentially, the GFR acts as an onsite, real-time, risk manager and insurance agent for the United States government.

As aircraft, especially highly instrumented and unique flight test aircraft, are quite expensive, contractors won't fly them until a GFR assumes financial liability for the aircraft. The job sounded like a great opportunity especially when combined with the experimental flight test of a new aircraft. I conveyed my interest to the AEDO detailer and was selected for the position.

The SH was designed to have common systems with the original (legacy) F/A-18 C/D aircraft to reduce development risk and cost. Improvements in the aircraft made it more capable and valuable to the fleet carrier airwings. The SH is a larger aircraft than the C/D with more fuel (translating to greater mission range or endurance), 2 more weapon stations (the SH has eleven stations) to carry a greater and more diverse ordnance load, a much greater maximum takeoff weight (66,000 lbs) enabling it to carry a larger load and a larger carrier bring-back weight (9000 lbs) adding tactical flexibility and safety with more fuel available for landing. The SH was designed to simultaneously carry Sidewinder, Advanced Medium Range Air-to-Air Missiles (AMRAAM) and HARM antiradiation missiles and Air to Ground (A/G) ordnance. That loadout was a dream come true for strike crews as it would allow them to defend their aircraft during an attack mission. It was the aircraft many wished the Navy had purchased with the original Hornet.

The Super Hornet Integrated Test Team (ITT) at PAX was comprised of contractors and government employees working side by side in a collaborative manner to facilitate and expedite flight testing[274]. The MCAIR ITT Director was test pilot, Pete Pilcher[275]. He was a quiet,

---

[274] The 434 ITT personnel included 300 from MCAIR, 26 from Northrup Grumman, 8 from GE and 100 from the U.S. Navy. Aviation Week & Space Technology, 20 January 1997.

[275] Pete was a Navy test pilot with many years of service with MCAIR; https://www.military.com/daily-news/opinions/2021/02/22/facing-shortfall-fleet-navys-workhorse-fighter-jet-deserves-our-support.html

calming influence for the team as he never got rattled when problems arose. Pete reflected his vast experience as a test pilot with concern but steadfast determination when presented with a challenge. Commander 'Birt' Wirt was the government lead for the SH flight test effort. He was an experienced navy test pilot with a strong A-7E and F/A-18 weapon separation background. Birt was more excitable than Pete by nature. He had a passion to understand the in-depth details of aircraft technical problems. Trying to understand all the technical details for an aircraft like the SH was mind-boggling. The ITT occupied and flew from the Hazelrigg Hangar, originally built for the A-12 program.

Before discussing the SH flight test program, it is appropriate to acknowledge all the team members that worked so diligently. I was able to see only a small portion of the work that occurred behind the scenes. Numerous awards were won by the ITT and the Super Hornet acquisition team. The performance of the SH in the fleet speaks for itself. The aircraft serves as the power projection backbone for the U.S. Navy aircraft carriers. For those interested in a more thorough discussion of the SH aircraft, the flight test program and its operational employment, I highly recommend Brad Elward's book[276].

Wednesday 29 November 1995 The first flight of the SH was conducted out of MCAIR facilities at St Louis Lambert International airport with the SH Chief Test Pilot (CTP), Fred Madenwald[277]. I was at MCAIR on Hornet business but made time as the incoming GFR to watch the first flight. There was a crowd of contractors, government representatives and civilian spectators assembled to watch the first takeoff. Fred made a comment after the flight suggesting the aircraft was ready for the carrier. While the comment sounded premature, it reflected the maturity of the flying qualities of the aircraft. Few changes were made to the flight control laws before the SH made its first landing aboard an aircraft carrier in early 1997. That is an amazing accomplishment and reflects the extensive hours of hard work by test pilots and the flight control engineers prior to the first flight.

---

[276] B. Elward, *The Boeing F/A-18E/F Super Hornet & EA-18G Growler; A Developmental and Operational History*, (Atglen, PA: Schiffer, 2012)
[277] Fred was an exceptional man with a big heart, broad smile and gentle demeanor. He was an excellent leader for the test program.
http://thetartanterror.blogspot.com/2012/10/frederic-amadenwald-iii-1951-2012.html

Thursday 1 February 1996 I checked into the tactical aircraft, flight test squadron, Strike and the ITT. Strike – radio callsign Salty Dog – flew F-14 Tomcats, F/A-18 Hornets and Super Hornets, A-6E Intruders, EA-6B Prowlers, AV-8B Harriers and the T-45 Goshawk aircraft. As there was great diversity in the aircraft, aircrew and projects, the atmosphere was unlike my past squadrons. There were aircrew going in many different directions with various deadlines mandated by external agencies. Everyone was perpetually busy with time critical testing. While the aircrew respected and appreciated one another, we didn't get as many chances to really get to know one another like in VX-5 or a fleet squadron.

Upon arrival at the ITT, there was a great deal to learn. I attended a week-long class in Arizona that explained the contractual importance of the GFR as the contracting officer's onsite representative. In our case, the U.S. government had signed a contract for MCAIR to deliver seven flight test aircraft and perform test flights over a three-year period for $4.3B. A great deal was at stake in our efforts. The flight test program was designed to ensure the aircraft was mature enough to successfully complete OPEVAL with VX-9. Each of the seven, highly instrumented, SH flight test aircraft were valued at over $100M. A loss of any of the aircraft could have compromised the entire flight test program. It simply would have taken too long to replace one of the instrumented aircraft. The flight test program needed to be executed in a rather flawless manner. Minus early growing pains and unpredictable aircraft anomalies, the program was well executed.

Before a new aircraft enters fleet service, many items need testing throughout the aircraft's approved flight envelope. The flight envelope means from the slowest speed to the highest speed (KIAS/Mach number), from 50 feet AGL to its maximum altitude attainable (service ceiling of 50,000 feet MSL), and from negative 3Gs to positive 7.5Gs. The envelope was expanded in a very careful and deliberate manner to increase the allowable flight regions quickly, yet safely. The following items represent a short (yet not comprehensive) summary of the routine items checked in flight test;

- Performance; rates of climb, acceleration, fuel consumption, range, and endurance with a variety of external store loadouts
- Loads testing; verifies that the aircraft can be safely flown through its flight envelope without structural concerns with all designated weapon configurations
- Flying qualities of the aircraft; all configurations, external stores, speeds and accelerations
- Flutter; no dangerous coupling of bending and torsional modes throughout the aircraft's flight envelope
- Stalls and spins; acceptable behavior when the wing is stalled and/or the aircraft is in out-of-controlled flight. This was critical as tactical aircraft routinely fly in this region
- Engine testing; smooth, predictable performance regardless of throttle movements; airstarts
- Takeoff and landing distances, speeds, crosswind performance to limits
- Weapons Separation to ensure release is predictable and safe throughout the designed envelope for a variety of weapons
- Crew Interface issues to ensure the crew can effectively and efficiently control the aircraft
- Aircraft carrier suitability to include arrested landings and catapult takeoffs with weapons
- Night flight performance from an aircrew interoperability perspective

To complete all the required testing in the most efficient manner required multiple aircraft. The ITT flew seven SH aircraft, five single seat E aircraft and two dual seat F aircraft. Each aircraft had a primary focus for testing. That testing required unique (and expensive) instrumentation in each aircraft. One aircraft often couldn't complete the tasking of another aircraft due to the embedded instrumentation. The aircraft primary testing assignments are listed below.

E1 was flight envelope expansion and flutter testing, E2 was the propulsion and performance test aircraft, E3 performed loads testing, E4 was the high AOA, spin and Out Of Control (OOC) flight aircraft sporting a high contrast orange and white paint job, E5 was a weapon separation aircraft, F1 was the carrier suitability aircraft and assisted in wing drop testing[278], F2 was another weapon separation aircraft with side stick controllers in the aft cockpit and also assisted in wing drop testing.

Each aircraft had a Test Conductor (TC) that led his/her team of specialist engineers for their specific area of testing. The TC kept track of the aircraft configuration and maintenance, the current flight clearance, the test plan status and helped to develop the test cards for each flight. He/she was also the focal point in the flight test control room talking to the test pilot during testing.

The situation in the control room was like what we see for a space launch although on a smaller scale. The TC job was a demanding position that required exceptional people skills along with technical excellence. The stress of schedule, dealing with unexpected challenges, while completing experimental test points that posed risk to the test aircrew and aircraft was a daily reality.

When the test team wasn't briefing a test flight, conducting flight tests or debriefing a flight, they were preparing flight clearances or test cards for the next flight. The TCs for the ITT follow; E1 was Tim, E2 was Les, E3 was Donna, E4 was Dave, E5 was Joan, F1 was Howard and F2 was Mike. There were others that filled in if a TC was sick, on leave or otherwise not available. The additional TCs were Forest, Jim and Kirk.

Naturally, there were teams of Flight Test Engineers (FTE), technicians, maintenance and administrative staff inside the ITT supporting the TCs. The ITT were also supported by off-site contractor and government personnel from all over the USA. There were various design and technology teams that were intimately involved in the activities. Their work led to various aircraft design improvements. The maintenance team were kept busy with routine items as well as unique tasks that had never previously been completed. The flight test teams couldn't keep the aircraft in the air and conduct the testing successfully without the support of the engineers and maintenance teams. It was an ongoing effort that required intensive focus with many members putting

---

[278] Wing drop testing responded to a discovered aircraft anomaly and is explained in later pages.

in 18 hours a day for extended periods of time. Their coolness under fire was inspiring to witness.

There were various flight test experts that would review test plans and keep track of discrepancies; Ron, Bernie, Chris, Rich and John come to mind. Various engineering teams around the hangar digested the flight test data to determine the next course to follow. Bill was the flutter team leader and meticulous in his work to keep the crews safe[279]. In experimental flight test, the engineers examined the real-time data from various instrumentation sources such as load revealing strain gauges or vibration sensors to verify that more demanding flight test could proceed.

While any of the test pilots could reasonably be expected to conduct any test, it was more efficient to have each pilot focus their contributions in a particular area. The MCAIR/Boeing test pilots that flew for the ITT were the CTP Fred (retired USMC pilot), Jim 'Sandy' Sandberg[280] (retired USMC pilot), Dave Desmond (prior USMC pilot), Phil 'Broadway' Pirozzi (prior Navy F/A-18 pilot) and later, Ricardo Traven (prior Canadian Air Force F/A-18 pilot). The Navy test pilots were CDR Rob 'Knockers' Niewoehner (F-14s), LCDR Tom 'Gurns' Gurney (F/A-18s), LCDR Dave 'Decoy' Dunaway (F/A-18s), Lieutenant Tom 'Corn' Hole (F/A-18s), Lieutenant Frank Morley[281] (F/A-18s) and later, Lieutenant Ken Hamm (F/A-18s), Lieutenant Matt 'Ho' Tysler (F/A-18s) and Lieutenant Mike 'Sting' Wallace[282] (F/A-18s). The WSOs on the team were Lieutenant Bryan 'Chum' Herdlick[283] (F-14 RIO and fellow SH GFR), USMC Major Matt 'Jams' Shihadeh (A-6E BN and F/A-18D WSO), Lieutenant Neil 'Lester' Woodward (A-6E BN) and myself. There weren't any contractor WSOs at the ITT.

---

[279] If the aircraft movements are not well damped throughout the flight envelope, accidents can occur.
[280] Sandy with YF-23; https://theaviationgeekclub.com/join-test-pilot-jim-sandberg-for-yf-23-pav-2-gray-ghost-walk-around/
[281] Story with Corn, Frank and Gurns; https://www.history.navy.mil/content/dam/nhhc/research/histories/naval-aviation/Naval%20Aviation%20News/1990/1997/may-june/pilots.pdf
[282] 'Sting' earned his callsign as he is an excellent bass guitarist and sings lead vocals like the musician, Sting.
[283] Chum was exceptional in all that he did; https://www.ndia.org/events/2020/10/20/1pst-precision-strike-technology-symposium/speakers-classified/bryan-e-herdlick-phd

Fred started flying E-1 at PAX to expand the safe flight envelope. He later flew a large portion of the E4 spin and Out Of Control (OOC) flight tests and helped define and solve the wing drop problem. Sandy flew many of the E2 engine tests. Dave flew most of the E3 loads tests. Day after day he had flights (often more than one a day) full of 7.5G data points. I don't know how his body held up during all the loads. Broadway focused on weapon separation tests. Ricardo worked on finding a solution for wing-drop then completed E4 departure testing near the end of the flight test program.

Knockers started flying E1, E2 then E4 for departure testing. He discovered the wind-drop problem and had many flights to define and solve the problem. Gurns and Frank flew F1 for carrier suitability testing including the first flights from the USS Stennis. Decoy, Corn, Ken and Ho primarily flew E5 and F2 in weapon separation tests.

The WSOs weren't initially flying the SH aircraft as management thought a minimum crew approach was more appropriate for safety reasons. The WSOs started flying after my first SH flight in September 1996. Our flights were initially low risk without weapon separation. This changed as confidence in the aircraft grew.

The ITT Flight Operations and Radio Room was an activity center for flight test. Jim was the lead flight support specialist that set up the room and managed the staff. Navy civilian employee, Scotty, ensured the correct support was available for each flight including test, chase and tanker aircraft, test ranges and ground support personnel. Bob managed the radios with his ATC background as a retired Navy Senior Chief. He was in touch with each flight test aircraft while airborne in the event they needed anything. This was quite a task when several of the seven flight test aircraft were flying simultaneously.

Billy was the MCAIR/Boeing maintenance manager for the SH aircraft. He was a tough yet poised, retired Navy Master Chief that had great experience in keeping aircraft flying. Billy had many military and contractor connections and brought skilled mechanics with him to the ITT. His team loved working for him and nothing was too difficult for them. The team was rather small to maintain the seven aircraft as modifications and repair of instrumentation seemed to be perpetual tasks. There were various specialist mechanics that performed maintenance and upgrades on the aircraft but always under the approval of Billy and his team. When presented with a significant problem (there were thousands during the three years of flight test), Billy managed the problem in a cool, professional manner and delivered mission ready aircraft without getting anxious or rattled.

I was fortunate to have a more experienced officer, Navy A-4 pilot, BJ, serving with me as a GFR for the first six months of flight test. We initially shared the role of the GFR. We initially reviewed volumes of MCAIR's safety and maintenance procedures. There was a lot to read and understand. It was amazing that the MCAIR team could follow the vast scope of their own procedures but many of the employees had prior military aviation experience. As the ITT had both government and contractor aircrew flying the test aircraft, we reviewed and maintained records for all aircrew ensuring their currency and competency to fly.

BJ and I attended an initial barrage of meetings upon arrival at the ITT. There were many things that everyone needed to know and agree with before we started flight operations. We were perpetually reading test plans, reviewing flight clearances and keeping up to date with the latest issues or discrepancies that were discovered on each flight. In general, the SH aircraft performed superbly. The aircraft software identified its own maintenance problems by setting onboard codes that were downloaded by the maintenance personnel after the flight. Flight after flight, the SH aircraft came back after very demanding flight testing with no discrepancies or 'popped codes'. While a bit shocking and mysterious at first, we became accustomed to the SH exceeding our expectations.

To conduct flight test in accordance with Navy procedures, a flight clearance outlining the aircraft configuration and the authorized flight maneuvers was required. Historically, AIR530 served as the NAVAIR flight clearance approving authority. The GFRs could not sign the liability release for an aircraft without a NAVAIR approved flight clearance. Birt fully supported that crosscheck and expected the GFRs to be the final checker before flight. He did not want to violate the trust NAVAIR had placed with the ITT. As the test program accelerated, we needed a more responsive flight clearance process. Captain 'Jeffrey'

Wieringa[284] assisted the ITT and NAVAIR by publishing a booklet that mapped out the old and new flight clearance processes. Flight clearances started arriving via email and before the test teams needed them.

The ITT had a database that kept track of all the discrepancies discovered during flight test. Standard practice with flight test was to categorize the problems by their severity and plans were put in place to resolve the issues. The problems ranged from minor ones that could be easily fixed to significant problems that required extensive flight time and engineering analysis to develop a solution.

The entries that follow provide mere glimpses into the flight test of the Super Hornet. Things started out slow as we awaited aircraft to arrive from St. Louis but ramped up quickly once all seven aircraft were present.

Wednesday 14 February Gurns flew aircraft E-1 from St Louis to PAX. It was a great day at the ITT! Our flight test program could finally commence.

Monday 19 February Aircraft E-2 arrived at PAX. A second aircraft was welcome as more test teams wanted to get busy.

Saturday 24 February NFO Jimmy D died in an accident while flying on a routine training mission in an EA-6B launched from the USS Kitty Hawk. He was in the class ahead of me at TPS and had provided great encouragement. Jimmy D was another amazing guy that died all too soon and is greatly missed by his friends[285].

Monday 4 March Knockers discovered an aircraft anomaly nicknamed 'wing drop' during the seventh SH test flight. He was in an accelerated turn in E1 when suddenly and without his input to the control

---

[284] Jeffrey and I first met when VA-115's three aircraft cross-decked to the USS Kitty Hawk in 1987. Admiral Wieringa enjoyed an exceptional career as an Intruder pilot, test pilot, Program Manager and Executive Leader; https://www.navy.mil/DesktopModules/ArticleCS/Print.aspx?PortalId=1&ModuleId=692&Article=2235973

[285] Jimmy D died while flying an EA6B aircraft from the USS Kitty Hawk. https://usnamemorialhall.org/index.php/JAMES_M._DEE,_LCDR,_USN

stick, the aircraft rolled up to 60°. Aircraft should not autonomously change flight paths without the pilot's input to do so. In close formation with another aircraft, an uncommanded roll could result in the loss of both aircraft and their crews. If a pilot were about to fire his gun or release a bomb and the aircraft did an uncommanded roll, targets could be missed. Wing-drop initially caught everyone by surprise as the SH was performing so well in all areas of its flight envelope.

Knockers flew an instrumented aircraft back to the Mach number and AOA where wing-drop occurred and reproduced the problem. The engineers were puzzled as they weren't sure what caused the roll. As engineers back in St Louis started looking at data and possible causes, flight test engineers were tasked with more precisely defining when the problem was occurring. We flew flight after flight defining the Mach, KIAS, altitude and AOA combinations where the problem occurred. Testing established the problem in flight conditions defined by 0.8-0.9 Mach and 8-10° AOA.

There was an initial thought that the problem could be solved with a software modification to control the leading or trailing edge surfaces of the wing. These changes solved much of the problem, but the final solution turned out to be more complicated than initially imagined. The solution took years and countless hours of computer simulation, wind tunnel analysis and flight time to define and implement. Critics of the SH program in Washington, D.C. tried to use the problem to kill funding for the aircraft. Fortunately, the aircraft, the ITT and MCAIR/Boeing survived the challenge.

Friday 15 March Intruder test pilot, Tim and I had my first refresher flight in a A-6E. It was great to be back in the aircraft as I hadn't flown one since a flight with Snow at VX-5 in December 1992.

Tuesday 19 March I needed a NATOPS requalification in the Intruder and VA-42 at NAS Oceana was too busy to fit me in. That required a trip across the country to Whidbey Island. I completed the checkride over several days with VA-128.

Tuesday 26 March Training for the SH aircrew was held as a group in the MCAIR spaces in St Louis over a few days. Several of the

SH pilots had been flying the simulators and more recently the aircraft so they knew it well. The rest of us were trying to come up to speed. The training provided an overview of the differences in the aircraft from the legacy Hornet. We were issued an interim flight manual that was the precursor to the NATOPS manual for the aircraft.

Friday 12 April Fred completed the first supersonic flight in E1 achieving M1.1 on a 3-hour flight. No anomalies were found with the aircraft.

Saturday 13 April Fred expanded the flight test envelope in E1 by achieving M1.52. Once again, no anomalies were found. The flight test team were encouraged by the quick and great success!

Wednesday 22 May E2 completed a five-hour flight assisted by inflight refueling. The SH reliability was impressive and we loved it.

Tuesday 4 June Birt and I flew in a F/A-18B chase aircraft supporting SH flight test. It was great to be part of the program and flying formation with a SH. Birt delayed flying early in the program as there was so much to do in his role on the ground. He was very proud to be watching the SH in flight. Inflight refueling from a Marine Corps C-130 aircraft allowed us to extend our flight time to 2.9 hours.

Monday 17 June Corn and I enjoyed a SH chase mission in a F/A-18B. He was an easy-going guy with a youthful, enthusiastic manner as he loved his work! Corn, like most Hornet pilots, let me fly the aircraft from the rear cockpit while in a loose formation. Inflight tanking extended our flight time to 3.2 hours.

Wednesday 19 June MCAIR Test Pilot Jeff Crutchfield died in a F/A-18C aircraft near St Louis, MO[286] while practicing for an upcoming airshow in Europe. He had been the instructor pilot for my first flight in a Hornet at TPS. Crutch arranged for my first NVG flight in a Hornet at China Lake while we were on our west coast field trip. He traveled with me to Italy for my DT-II in the Tornado. Crutch worked in F/A-18 software development at the Naval Weapons Center, China Lake while I

---

[286] Jeff's passing was a tragic event for many; https://www.latimes.com/archives/la-xpm-1996-06-20-mn-16723-story.html

was at VX-5. He was a leader in the use of NVGs in the Hornet, a superstar naval test pilot and his loss remains a tragedy for those that were fortunate to have known him.

Tuesday 2 July Knockers and I shared a SH chase flight in an F/A-18B. He was articulate yet often quiet in the aircraft unless something needed to be said.

Monday 29 July TPS classmate, Spot and I had a day trainer flight in an F/A-18B. We did acrobatics to build up our G endurance. It was our last flight together[287].

Monday 19 August Marine Hornet test pilot, callsign Knife[288] and I shared our first flight together while conducting a familiarization of the CIT system in a F/A-18D. He and were neighbors in Calvert Country and spent time together socially. This was the first of many enjoyable flights with Knife as we flew together for ATARS flight testing[289].

Thursday 22 August Sandy flew E-4 from St Louis to PAX. As he approached from the west, Jeffrey and I took off in an F/A-18B for us to capture photos of the beautiful E-4 orange and white paint job with PAX in the background.

---

[287] At the age of 59, Spot very sadly died of prostate cancer in 2019 following a great career in naval aviation. His spouse, Sheila, successfully lobbied Congress for an investigation into the premature, cancer deaths of military aircrew.
https://myemail.constantcontact.com/James-William-Galanie-1959---2019.html?soid=1107630229941&aid=1EKdONFMVRM
[288] Knife's callsign reflected his quick, clever and powerfully penetrating wit.
[289] Knife is exceptional in everything he does. He manufactures beautiful furniture from solid beams of timber, renovates houses, works as a consulting engineer and flies commercially in the USA.

Thursday 29 August I visited NAWC Lakehurst, NJ to inspect their facilities and evaluate their preparation for the upcoming detachment with the F1 test team. Carrier suitability tests were scheduled to commence in mid-September in preparation for initial sea trials in January. The Lakehurst staff provided a solid overview of their capacity for testing. Lakehurst unfortunately had a FOD problem with various debris around the runways that needed to be remedied prior to the detachment.

Friday 13 September Horse and I shared my last flight in an Intruder. The flight was a FCF 'A' Profile that was required after removing and reinstalling engines. The flight was bitter-sweet after flying Intruders around the world.

Friday 20 September Jeffrey and I chased E1 around the skies in a F/A-18B for 3.4 more hours of flutter testing. I was looking forward to my first flight in F1 on Monday.

Monday 23 September Prior to today's flight with Corn, ITT management would not approve of a NFO in the backseat of the F aircraft during flight test. The usual rationale was that the flying was too dangerous and they wanted to minimize the risk to aircrew. As time passed, it was advantageous to get a WSO's perspective. F-1 was being repositioned to NAS Lakehurst, NJ for arrested landing testing. The flight was short and the flight profile benign. I needed transportation to

Lakehurst to supervise flight operations and sign aircraft release forms and was scheduled to fly F1.

At this point, I knew how Hornets handled having controlled them from the back seat on numerous occasions while at TPS, VX-5 and at Strike. The Super Hornet is heavier than its predecessor with the ability to carry more ordnance and fuel. I partially expected the new aircraft to be slower on takeoff than the legacy Hornet considering that extra weight. I was delightfully mistaken. When the afterburner kicked in, I was smiling from ear to ear. From the first takeoff roll, I knew that the Super Hornet was going to be a game changer for naval aviation.

Once airborne, I recognized the familiar Hornet flying qualities yet they felt refined. Once we were in level flight, Corn asked if I'd like to control the aircraft. I took over the controls and evaluated the response of the aircraft to small inputs maintaining altitude and heading. While it felt like a Hornet, it felt smoother than the legacy Hornet. As we approached Lakehurst, Corn resumed control of the aircraft. The approach to landing was smooth with the engines very responsive to any changes commanded. I was impressed and we hadn't done much in a forty-minute flight.

The Lakehurst detachment was a success although it presented challenges. We had concern for the cleanliness of the runway testing facilities after my initial survey. The last thing we needed was engine FOD. Our testing generated shards of steel that were being thrown around the runway. We instituted periodic and frequent FOD walkdowns to ensure that we didn't lose an engine and precious test time due to damage. Our efforts proved to be effective and the testing was successfully completed.

Saturday 5 October Gurns and I returned F1 back to PAX after testing at Lakehurst. He was a calm, mature, soft-spoken gentleman that chose his words carefully yet loved a laugh. While flying the aircraft, Gurns was smooth. He had many hours flying the aircraft at Lakehurst and was pleased with progress to date.

October Test pilots, Mort[290], Greg and I flew three flights in support of the T-45 Cockpit 21[291] evaluation. These flights were especially gratifying as the pilots had me fly the aircraft in the traffic pattern at PAX. The pilots were evaluating how well the HUD and other cockpit displays presented flight data during landing practice. After each landing, the pilot applied full power, started the climb away from the runway and turned the controls over to me. This enabled them to be heads down writing debrief notes for the approach and landing. They took back control of the aircraft on the base legs. Mort and Greg appreciated the help as it enabled them to provide better feedback on their assessment of the aircraft.

October E4 had completed modifications and initial check outs at PAX and was ready to start the high AOA testing. As the aircraft was dedicated to spin and OOC flight testing, it was painted orange and white to help the ground-based cameras record the aircraft's motion more clearly. As the aircraft and pilot could potentially get into a flight regime that would not allow a pilot-controlled recovery, an Emergency Recovery System (ERS) was installed on the aircraft. The ERS had electrically driven backup hydraulic pumps with independent batteries and a pilot-deployable parachute mounted between the tails. Once deployed, the parachute needed to be severed prior to returning for landing as it generated unacceptable drag.

There was considerable, tense debate regarding the wisdom of inflight testing of the ERS parachute to validate its functionality. There was concern that if the parachute deployed and couldn't be released, we could lose the aircraft severely threatening the test program. On the other hand, we wanted to ensure that the pilot possessed a validated tool to recover the aircraft if the flight controls were ineffective. After extensive briefings and risk analysis, it was decided to flight test the parachute.

Wednesday 9 October Fred flew E-4 into a safe zone over the Chesapeake Bay in the event the parachute failed to release yet in a location where it could be retrieved if it did. In the first and only airborne

---

[290] In June 2001, Mort assumed command of Strike; https://www.NAVAIR.navy.mil/node/3426
[291] Cockpit 21 introduced digital cockpit displays to enhance training; https://janes.migavia.com/inter/boeing-bae/t-45.html

test of the parachute system, Fred deployed the parachute. A chase pilot and Navy photographer, Vernon Pugh, were in a perfect position to gather a stunning picture. Fred conducted flying qualities testing over the fifteen seconds after inflation to determine the parachute's impact on the aircraft. Once satisfied with the feel of flying with the parachute, Fred jettisoned the parachute with anchor bolts exploding as designed. Fred landed the aircraft back at PAX and the parachute was recovered from the water as planned. The system had worked as designed and the high AOA/spin/OOC flight testing commenced.

Saturday 16 November Knockers was completing supersonic testing in E1 when a sixth stage stator blade (stationary airfoil) of the F414-GE-400 engine failed damaging the engine (Springsteen & Bailey, 1998, pp. 19). Knockers reduced power on the engine and landed the aircraft safely at PAX. The engine was shipped off to GE for analysis. Following many discussions and meetings, a decision was made to suspend flight operations until a clear explanation of the failure was found.

It was determined after an in-depth investigation that metal fatigue caused the stator failure. Other aircraft were inspected and more fatigued stators were discovered. It was proposed that the stator design had a problem rather than the flight test causing the failure. Through lab testing, it was determined that a small modification in the design of the stator caused the fatigue and other parts were available to retrofit into the engines. All the aircraft were grounded for three weeks before limited flight operations resumed. The aircraft engines needed modification before full flight testing could resume. Priority was given to F1 to continue flights in preparation for initial carrier qualification in January. The ITT maintenance team worked tirelessly throughout December to get all the aircraft back to flight status. Once we started flying again, the ITT worked three of four Saturdays a month to regain lost time.

Thursday 21 November Requirements officers from the Pentagon hosted a F/A-18F future requirements briefing. It was noted that senior officers at the Pentagon including CNO Johnson, a F-14 pilot, were in favor of an upgraded F aircraft onboard the carriers. A revised ORD was in work to cover future technical developments. All weather, precision strike using Joint Direct Attack Munition (JDAM)[292] aided by the AESA EXP 4/5 radar modes and the ATFLIR was presented. Funding plans and timelines were provided that included AESA, ATFLIR, EXP 4/5, JHMCS, decoupled cockpits, upgraded displays, aft seat lasing, independent map formats, independent and simultaneous radios transmissions and Satellite Communication (SATCOM) to support Tactical Air Coordinator (TAC) and FAC roles. Plans were in work for an Operational Advisory Group (OAG) process to evaluate tradeoffs in future developments. The initial Fs to hit the fleet would be like the Marine Corps F/A-18D with upgrades to follow.

Early December I was fortunate to complete the ground syllabus and three flights to obtain a F-14A/B NATOPS qualification at VF-101, the Tomcat FRS at NAS Oceana. The FRS syllabus was straightforward. Over the next twenty-two months, I enjoyed flying the Tomcat at Strike in support of various flight test projects.

---

[292] JDAMs could revolutionize aerial bombing; https://www.navy.mil/Resources/Fact-Files/Display-FactFiles/Article/2166820/joint-direct-attack-munition-jdam/

F-14 Tomcat, Naval Flight Test Museum, NAS Patuxent River, MD.
Photo by David Maybury

December Wind tunnel and computer predictions for the original SH design showed potential problems with weapon separation in certain flight conditions. Specifically, bombs could collide with one another or the aircraft after release. A great deal of discussion and debate resulted in a decision to angle the aircraft's under-wing pylons 4° outboard from streamlined at the leading edge. Simulations showed that this modification would prevent the weapon release problems. The seven flight test aircraft were modified with angled pylons with the first modifications commencing in December. The modifications were extensive requiring considerable time to remove the old pylons, redrill the wings for fasteners and reinstalling the pylons. Aircraft were available for this maintenance while completing the engine stator repairs.

# Chapter 15

## Super Hornet Testing 1997-1998

*Most people might imagine that when a pilot released a bomb or missile from the aircraft, it would drop gracefully away from the airframe and provide little risk to the aircrew. This is not always the case.*

Monday 6 January 1997 As the Super Hornet aircraft returned to service after engine repairs, FCFs were required on the aircraft.

Wednesday 8 January Tomcat test pilot, callsign Dex and I flew a night FCLP session in a Tomcat. Dex's professionalism and skill made me feel right at home in the aircraft.

Tuesday 14 January I arrived early at the ITT to support Frank's 7am brief for a flight in F1. Dex and I had another night FCLP session in the Tomcat with a 6pm takeoff making it a long day.

Wednesday 15 January Gurns and I flew in the FCLP pattern in F1. It was great to witness him fly numerous perfect approaches in the aircraft. I was impressed by the view of the ball from the back seat as it was unobstructed all the way to touchdown. We took fuel from a C-130 to extend our mission, perform engine anti-ice tests and to check telemetry with our team aboard the USS Stennis. They reported a weak signal that was concerning. Gurns and the jet were ready for the carrier but we planned to reassess the telemetry.

Friday 17 January Frank and I flew in the FLCP pattern in F1. The winds were gusty but the approaches were flawless. We flew to Norfolk to recheck the telemetry. The signal was strong and everything was ready for the initial sea trials.

Saturday 18 January Frank was scheduled to make the first landing with F1 aboard the USS Stennis but weather at the carrier was rain with a 1500-foot ceiling, temperature at 35°F and winds to 40 knots. The ship was positioned 60 miles east of North Carolina. There was concern inside the ITT that perhaps this wasn't the best weather conditions to land aboard a carrier for the first time. Once the risks were

273

thoroughly discussed, it was decided to launch the aircraft as Frank could always return if the conditions were unsuitable for the testing.

Frank reported that the aircraft handled like a dream as he approached for landing behind the carrier. This was due to the countless hours of effort by engineers and test pilots to perfect the flying qualities for the task. Lesson learned; put the hard work in at the start then success can be achieved despite the demanding conditions. Lieutenant Frank Morley[293] successfully landed F1 aboard the USS John C. Stennis to commence sea trials. The cover of this book shows the approach for that landing.

After the first landing, flight operations were suspended for high winds. The winds decreased later that day and Gurns piloted F1 during the first catapult shot. He had been the chase pilot in a F/A-18B and landed aboard Stennis with photographer, Randy Hepp. Gurns and Frank stayed aboard Stennis for six days making sixty-one daytime, carrier catapult launches and landings. The pilots examined landing approach flying qualities, dual and single engine handling qualities as well as aircraft trim and crosswind effects coming off the catapults[294]. The successful completion of operations aboard the Stennis provided a huge morale boost for the entire ITT.

Monday 20 January Tomcat test pilot Pistol and I flew a night FCLP session in the F-14. He was a relaxed, confident pilot and enjoyable to fly with.

Aviation Week & Space magazine reported that the F/A-18F was due to start replacing F-14 Tomcats in 2001. SH testing had achieved flight to 49500ft MSL, M1.54, +7G, -1.7G, +57° and -39° AOA; flutter neared completion, loads testing had started, the spin chute had been tested. A purchase of 670 E's and 330 Fs was planned[295].

---

[293] Vice Admiral Frank Morley's success continued well past his work with the ITT; https://www.navy.mil/Leadership/Flag-Officer-Biographies/BioDisplay/Article/2236288/vice-admiral-francis-morley/
[294] Aviation Week & Space Technology, January 27, 1997, pg 28.
[295] Aviation Week & Space Technology/January 20, 1997, pgs 54-59.

Wednesday 22 January F2, the second weapon separation aircraft, arrived at PAX. The aft cockpit was fitted with side-stick controllers like in F/A-18Ds instead of the stick and throttle in the back seat of F1. Meetings were held to review test plans for propelled air to ground weapon releases.

Tuesday 28 January Dex and I flew in a F-14 as the chase aircraft for a Tomcat performing Digital Flight Control System (DFCS)[296] flight test. DFCS replaced the traditional F-14 flight control system with a fly-by-wire system to alleviate various aircraft control problems and improve carrier approach behavior. The new system needed to be thoroughly tested throughout the aircraft flight envelope to ensure it didn't introduce any new problems.

Saturday 1 February E3 arrived at PAX completing the total of seven test aircraft at the ITT.

Monday 3 February Sandy and I flew a mission in F1. He was a quiet, reserved pilot that did everything methodically. After engine start and checks, we slowly taxied out to the runway for takeoff. Within ten minutes after takeoff, we were at altitude doing supersonic testing of the Environmental Control System (ECS) system at M1.45. We completed inflight refueling to extend our flight time and more supersonic testing. The aircraft was smooth, quiet and stable despite the high speed. I was able to test the TFLIR and radar in A/A and A/G modes with classified results passed onto the mission systems team.

The ITT added F-14 RIO, Chum, to serve as another GFR. He was a recent TPS graduate, passionate about Tomcats and a very smart character. Once his training was complete, we shared the load of GFR duties and covered for one another during periods of illness, training or leave.

Wednesday 19 February Broadway successfully completed the first external stores separation test in E-5 by dropping a 480-gallon empty fuel tank from 5000 feet MSL into the Chesapeake Bay.

---

[296] DFCS history; http://www.anft.net/f-14/f14-detail-dfcs.htm

**Wednesday 26 February** For the first time, a SH aircraft (E1) carried a combat representative loadout with three 480-gallon drop tanks, two MK-84 (2000lbs each) bombs, two HARM antiradiation missiles and two AIM-9 Sidewinder air-to-air missiles. The refueling aircraft had a Navy photographer take a picture of the event[297].

Twenty-three members of the ATARS flight test team were assembled for a photo in front of a F/A-18D with the pod attached. The team were a great bunch and we enjoyed working with them. It was very gratifying to go from design concept at NAVAIR to flight test at PAX.

**March** Weapon separation testing accelerated with the releases of MK-82 bombs, SLAM and Harpoon missiles. Weapon separation testing was always an exciting time in experimental flight test. Most people might imagine that when a pilot released a bomb or missile from the aircraft, it would drop gracefully away from the airframe and provide little risk to the aircrew. This is not always the case. The dynamic pressures and flow interferences that occur around a tactical jet aircraft at high speed can generate very unpredictable weapon releases[298]. Decoy, Broadway, Corn, Ken and Ho completed most of the weapon separation testing. WSOs were present in the rear cockpit of F2 for several of the later weapon separation flights.

**Friday 28 March** CNO Admiral Jay Johnson[299] and Deputy CNO (Air) Admiral McGinn[300] had their first flights in a Super Hornet. Gurns piloted from the front seat in F1 for both flights. CNO Johnson had been a fleet F-14 Tomcat pilot and was excited to fly in the Navy's newest fighter. The pre-flight brief was held in the center conference room with a dozen people present. I joined Gurns and the CNO as they walked out to the jet. My role was to assist the CNO with the F rear cockpit displays prior to engine start. After a few minutes of assistance, I climbed down from the LEX, Gurns started the engines and they taxied for takeoff. Both

---

[297] The image can be purchased here; http://hdwpro.com/f18-hornet-picture.html
[298] Some examples of weapon separation testing gone wrong can be viewed here. Photographer Randy 'Fireball' Hepp is in the backseat of the mishap A-4 aircraft; https://www.youtube.com/watch?v=Tynni6wpYZ0 and https://www.youtube.com/watch?v=fPTnmZ_HPAs
[299] CNO Johnson; https://www.history.navy.mil/browse-by-topic/people/chiefs-of-naval-operations/admiral-jay-l--johnson.html
[300] Admiral McGinn career overview; https://en.wikipedia.org/wiki/Dennis_V._McGinn

admirals returned impressed from their SH flights noting the aircraft was the future of naval aviation[301].

David discussing the new aft cockpit controls and displays with CNO Johnson as he prepared for his first flight in the Super Hornet.

Wednesday 9 April All available Strike NFOs came together in the morning to discuss the F rear cockpit as we considered the demands of a two-seat naval strike aircraft. As the F/A-18F was going to replace the Intruder and Tomcat as the high threat, precision strike asset for the carrier, the NFOs reviewed the design to meet the mission demands. The EA-6B ECMOs provided their perspective regarding the tools needed to complete their EW mission[302]. We developed a list of proposed changes to the aft cockpit and a few for the front cockpit (pilot push to talk ICS was one). I agreed to discuss the changes with Lyle at MCAIR to get an early feel for our suggestions.

Friday 25 April The ITT established a SH Deficiency Review Board (DRB) to review issues identified with the aircraft. In a contractual and legal sense, something was a 'deficiency' if the aircraft or system

---

[301] The CNO after the flight on page 13;
https://www.history.navy.mil/content/dam/nhhc/research/histories/naval-aviation/Naval%20Aviation%20News/1990/1998/july-august/review.pdf
[302] The F/A-18G Growler wasn't a topic of discussion yet, but the ECMO inputs helped the design suggestions.

does not conform to the aircraft specification and contractual requirements. Flight test aircrew are taught to identify issues with an aircraft then rank the issue in terms of severity. The test crews often document what they consider to be deficiencies, sometimes even on attributes that comply fully with the aircraft design specification documented from the aircraft contract. In this situation, a DRB meets to determine if the item identified should be corrected at government cost. If so, the government contracting officer negotiated with the contractor to determine a reasonable cost to remedy the issue.

After a previous flight in a F/A-18F, I noted that the touch-sensitive Up Front Control Display (UFCD) in the rear cockpit was difficult to read and use in its current location. It served as the primary data entry device for the aircrew. I wrote a deficiency on the UFCD as installed in the aft cockpit. The DRB determined that my deficiency was out of scope for the flight test program. I appreciated their perspective, yet it was worthy of future consideration. Many of the NFO comments were labelled out of scope as the test program contract didn't consider the carrier-based, F aircraft with a WSO. Fortunately, the NFO comments did eventually lead to changes in the SH Block II and Block III aircraft.

Tuesday 29 April Frank and I flew a F/A-18D as the chase aircraft for E-4 as it performed high AOA, spin and OOC flight testing. The initial E-4 pilots, Fred or Knockers, were very busy working through all the possible departure scenarios. On these flights, the E-4 pilot and his chase aircraft climbed to 40,000 feet to start testing. E-4 was slowed to stall speed and a flight control input was introduced. The pilot evaluated the aircraft's response to the input, recovered the aircraft and climbed back up to 40,000 feet to try another input. Fred or Knockers placed the aircraft in spins, evaluated its character in the spin as stable, accelerating or benign. They would evaluate the character of the aircraft in its response to recovery controls.

Getting into a spin isn't unusual while performing high angles of attack, low speed dogfighting with another aircraft. The original F/A-18A/B/C/D aircraft had some uncomfortable flying qualities after departure from controlled flight. One departure mode was nicknamed the 'Falling Leaf' and it resembled that motion. It required pilot training to recognize the character of that flight region and recover from it. During a

'Falling Leaf' departure, the aircraft remained in a stalled condition at high AOA and could not be recovered unless the pilot executed specific flight control inputs[303]. The ITT was hopeful that the software engineers had eliminated the possibility of similar flying qualities with the SH. They did a great job and no such flight regime existed!

Wednesday 7 May Broadway and I had our first flight together in F2 lasting 1.6 hours. He was in the class ahead of me at USNTPS and had flown with VX-4 at Point Mugu while I was at VX-5. We had limited interaction with one another before the ITT.

Thursday 8 May Decoy and I had a flight in F2, our first flight together since VX-5.

Monday 19 May A government employee, callsign Troll, joined us as an additional ITT GFR. He had F-4 experience as a naval aviator and was a jovial character. He became a valuable addition to our office, helping to keep up with the busy flight schedule and adding wisdom as events unfolded[304].

Tuesday 20 May Decoy and I had a 1.6-hour flight in F2. It was great for the ITT to have his OT perspective embedded in the SH flight test. Decoy could predict OT problems and was in regular contact with Zoil who led the SH OPEVAL. It was interesting how Vampires of the early nineties were working the SH development in the late nineties.

Wednesday 21 May Pistol and I flew the VR-1754 low level in a Tomcat for recurrency training. The aircraft was a sweet ride at 200 feet AGL and 420KGS with the wings swept back. The F-14 was a bigger, heavier aircraft but had low altitude flying qualities like the Tornado.

Wednesday 28 May Corn and I had a 1.6-hour flight in F2. While Corn was conducting test points in the front seat, I examined the resolution of the A/G radar in the back seat and provided feedback to the mission systems team.

---

[303] A Canadian pilot shares his story of a Falling Leaf here; https://www.youtube.com/watch?v=QSLmblzI_c4
[304] Troll served as the GFR during F-35 flight test program following the completion of the E/F flight test effort.

Friday 6 June Marine pilot Eli and I provided video chase in a F/A-18B for E-4 departures. I carried a Hi-8mm video camera in the back seat to film the departures. Both of our aircraft started at 40,000 feet with us orbiting around E-4 as it slowed to stall. As the E-4 pilot neared departure from controlled flight, I started recording while focused on the aircraft in a medium field of view. Eli flew a diving, circling pattern around E-4 while keeping a safe distance. Both aircraft were descending at rates up to 20,000 feet per minute. The maneuvers lasted up to a minute and we usually ended up in a 30° dive in the process. I tracked E-4 until the maneuver was completed and the recovery commenced. Maintaining the aircraft in the camera viewfinder throughout the maneuver was challenging but rewarding. It was like tracking a target during a bombing run but in this case the target and the camera were moving. Small movements in the handheld camera produced unsteady video.

The transition from level flight to the spiral dive was tricky. The flight maneuver was simple to execute but more difficult to execute well. The transitions to the dive were often bumpy and every little bump or acceleration made it more difficult to obtain good video. As my experience grew with recording these maneuvers, I better understood how to brief my pilots. In time, the video improved becoming impressive to watch.

Monday 23 June Dex and I had an afternoon Tomcat training flight scheduled. He asked about my preferences for the mission. I sensed that he wanted to show off his fleet airframe when he asked, "Have you been to our VMAX of Mach 1.77 yet?" I had not come anywhere close to that as other aircraft were slower. "That's it then. We'll go out over the ocean to the east in the supersonic corridor and get the bird up to maximum Mach". A normal training hop in a Tomcat would last about 1.5-1.7 hours. This flight lasted 1.0. You burn a lot of fuel when travelling that fast.

Once off the coast in the Warning area, we flew to the very end and turned around to have enough room to achieve our top speed and stay in the area. Dex added full afterburner and we started accelerating. Passing Mach 1.0 was a non-event. As we passed M1.5, I was surprised how smooth the aircraft was and how the acceleration was still strong. As we approached M1.77, Dex pulled the power back and we stabilized at that speed. The aircraft was smooth and quiet. Dex explained that the jet engines could take it faster but the windscreen aerodynamic heating limited the top speed. Dex slowed the jet to subsonic speeds and we returned to PAX.

Tuesday 24 June Hornet pilot Miggs[305] and I had a NVG refresher flight in an F/A-18D along the VR-1754 through the Allegheny Mountains west of PAX.

Thursday 26 June Knife and I flew a CIT test flight in a F/A-18D. It was another product that I worked on at NAVAIR then flew at Strike.

Saturday 28 June An article that highlighted the Super Hornet was published in the magazine, Naval Aviation News. JO2 Towler alias LT Al E. Ron did a great job convincing the reader that he had flown a SH[306].

August 97 MCAIR merged into Boeing making it a bigger and more diverse company. While the larger company had advantages, many employees grieved the loss of their original company name that was an American icon.[307]

Thursday 4 September We celebrated Pete Pilcher's 30th year working for MCAIR. He had a startling flight in an F-4 early in his career[308] and was part of the initial Hornet flight test team at PAX in the early 80's. Despite his vast experience, he was a soft-spoken and humble man.

---

[305] Miggs had a very successful career with the Navy; https://www.militarynews.com/norfolk-navy-flagship/oceana/news/zins-reaches-milestone-with-1-000th-trap/article_b06c449a-1ec0-53cb-892e-1b7b257d283c.html
[306] Super Hornet stories; https://www.history.navy.mil/content/dam/nhhc/research/histories/naval-aviation/Naval%20Aviation%20News/1990/1997/may-june/hornet.pdf
[307] MCAIR short history; https://www.britannica.com/topic/McDonnell-Douglas-Corporation
[308] A part problem generated an exciting start and end to the flight; https://www.youtube.com/watch?v=yA_QN1VdZPw

Monday 29 September I was fortunate to fly in the back seat of F-2 for an AMRAAM missile launch. Broadway was the pilot for the shot. For this flight, the engineers wanted the aircraft to reach a very specific Mach/airspeed/G load/altitude combination that was demanding from a weapon separation standpoint. To make things more difficult, the flight tolerances for the shot were tight; +/-0.02Mach, +/- 10KIAS in airspeed, +/- 200 feet in altitude and +/- 0.2G in loading. It would have been helpful if the release point had been in level flight but the jet wasn't predicted to get to the designated conditions while level. Once airborne, Broadway tried a couple of practice maneuvers to achieve the release parameter combination. The practice maneuvers confirmed that the aircraft required maneuvering to reach the release parameters. We needed to start the maneuver thousands of feet higher than the desired firing altitude and dive into the point to achieve the correct conditions. If that wasn't hard enough, the maneuver needed to be loaded to 5Gs. Broadway considered the maneuver required to hit the parameters and visualized a plan.

We made a practice run. Broadway started high and commenced a spiraling, accelerated descent that hit the airspeed/Mach/altitude/G load as desired. I was amazed! The engineers were impressed as well. Did Broadway get lucky or could he do that again? We did another practice run and he nailed the parameters again. The engineers felt confident that the plan would work. Broadway was instructed to set up for another run with the intention of weapon release. If the engineers felt that the aircraft wouldn't achieve the desired release point, the TC would call "knock it off" to stop the test. If everything looked good, Broadway would continue and release the weapon. He initiated the third run; it looked like the previous runs and a successful release occurred. That was one of the more demanding flight test sequences I'd ever experienced.

Friday 10 October Ho and I launched a Maverick missile during a test flight in F2 on our first flight together. Ho was one of the younger Navy pilots but had excellent skills.

Tuesday 14 October Boeing pilot Ricardo was finally given permission to fly the SH. We reviewed his paperwork in preparation for flying. He had been working on a solution to wing-drop due to externally managed security issues with his Canadian citizenship. He was anxious to start flight testing and we were excited to get him airborne[309].

Flight testing clearly defined where wing-drop occurred. Once we defined the problem, we needed to fix it. This is where science met art and craftsmanship. The aerodynamics group from the ITT worked with their counterparts at MCAIR, St Louis. They thought that tripping the air flow boundary layer on top of the wing might eliminate wing-drop. This could be accomplished with a strip of wood placed at a strategic location on the upper wing surface. Questions remained about the size, shape and location of the wood strip. The plan was to manufacture the strips, glue them to the top of the wing then fly the aircraft to determine their impact on wing-drop. If the problem was eliminated, we would fly other flights to confirm the impact of the strips on performance and flutter. If all areas checked out well, the fix would be manufactured out of a more durable material, tested and approved for production.

An Amish carpenter nicknamed 'Ziggy' near PAX was hired to manufacture strips of mahogany to engineer's specifications of length, width, height and shape[310]. The strips were glued into place on the upper surface of the wings at specific locations. The glue was amazing as it cured overnight and was ready for another flight in the morning. The cure had to be strong as the aircraft was flying near supersonic speeds on each flight and the velocity on top of the wing was very easily supersonic. There were various shapes tested. Isosceles and equilateral triangles were popular.

The SH aircraft with the Amish-made strips took off with the chase aircraft, flew out into the restricted area and to the region of the flight envelope where wing-drop occurred. A series of accelerated maneuvers was conducted to evaluate the new configuration in its ability to resolve the problem. If the configuration didn't show promise, the flight was terminated and the aircraft returned to the airfield. The engineers debriefed the test pilot with regards to his experience and another configuration was ordered for the aircraft. The aircraft mechanics

---

[309] Ricardo on aviation; https://globalnews.ca/video/3351746/boeing-super-hornet-chief-test-pilot-ricardo-traven
[310] Ziggy kept the SHs flying; https://www.baltimoresun.com/news/bs-xpm-1997-02-07-1997038112-story.html

got busy taking off the failed configuration and installed a new configuration.

In November, it was decided to use whatever tools were available to solve the wing-drop problem. As large condensation clouds often appeared on top of the wings during wing-drop testing, it was decided to record the tests with a video camera. The clouds could assist in understanding the air disturbance leading to the uncommanded rolls. The easiest and quickest method to record the events was to place an aircrew in the back seat of a F/A-18F with a hand-held Hi-8mm video camera.

When I first considered the task of video-taping the upper wing, it seemed rather simple. However, to achieve 8-10 degrees AOA at 0.8-0.9 Mach, the pilot needed to pull 7.5Gs in a turn. The challenge of filming the maneuver was compounded as the turn was not smooth but often bumpy in buffet. Imagine trying to keep a video camera steady and focused on a spot while driving down a bumpy dirt road or completing a 3G loop on a rollercoaster. The 7.5G loading magnified the challenge of keeping the camera steady during the testing.

Thursday 20 November Knockers and I flew our first wing-drop flight together in F1. The first test point was humorous from my perspective. As Knockers set up the aircraft for the first test point, I leaned forward while twisting my torso to optimize the viewing angle for the camera. As Knockers pulled on the 7.5Gs, my body and the camera fell forward under the load. I was pinned against the cockpit glare shield and could not lift my torso up against the aircraft acceleration. Fortunately, the test point only lasted about 15 seconds. I laughed while waiting for Knockers to relax the back stick pressure and release me. Knockers asked, "How did the video turned out?" I laughed and said, "Not very well. I misjudged the load factor. I'll try something different on the next point".

I considered my options for filming under 7.5Gs while holding the camera steady. I decided to sit up vertically in the ejection seat and create a rigid pedestal with my arm for the camera to rest on. If done properly, the pedestal was less vulnerable to the high G loading. I brought my right arm close into my body with the elbow at my stomach and forearm creating a vertical mount as the camera cradled in my palm. I

284

filmed over the left wing to capture the best field of view. I hit the camera record button before Knockers rolled into the next point. As the G-load increased to 7.5, my arm and the camera depressed but maintained its view on top of the left wing. The scene was quite spectacular with streams of condensed, moist air flowing over the wing. We performed 20-30 of those test points on a 2.5-hour wing-drop flight. I worked diligently to get the very best video on each test point. After each flight, I passed the video tape to the ITT engineers for analysis along with their other data.

Tuesday 9 December The SH aircraft surpassed the 2000 flight hour milestone with Ho at the controls firing an AMRAAM missile[311].

Thursday 11 – Saturday 13 December I recorded video during five wing-drop flights with Knockers and Fred in F1 totaling 7.8 hours over three consecutive days. As we repeatedly pulled 7.5Gs, it was a strenuous workout for us. Despite the effort, the solution we were seeking continued to elude us.

Tuesday 13 January 1998 Much of the SH had been tested by this point in the program. We needed to test the radar altimeter to ensure that it provided the pilot accurate information regardless of speed or altitude. Special flight test permissions were granted to fly the aircraft down to fifty feet above the water at a variety of speeds up to, but not including supersonic flight. As the aircraft was so low to the water with its added risks, I was invited to fly in the test. My task was to ensure that the pilot was flying safely and there were no unusual anomalies that could jeopardize the aircraft or crew. Sting[312] was the very capable, trustworthy pilot for the test. We had not flown together previously but his skills as an exceptional pilot were obvious from the preflight brief onwards. Once airborne, we commenced testing the radar altimeter and gradually lowered the altitude to fifty feet above the water. The aircraft was rock-steady at that altitude.

---

[311] https://boeing.mediaroom.com/1997-12-11-F-A-18E-F-Super-Hornet-surpasses-2-000-Flight-Hour-Mark. In 2002, Ho was the first pilot to reach 1000 hours in the SH aircraft; http://www.aero-news.net/index.cfm?do=main.textpost&id=d87cd8ee-a8d6-46bf-b1b8-8b5edda286cc
[312] Sting later flew for Boeing; https://www.boeing.com/news/frontiers/archive/2004/march/mainfeature2.html

Sunday 15 February The Washington Post released an article discussing the political risks to the SH program due to wing-drop[313]. There were factions in Washington D.C. that wanted to cancel the SH program and were looking for ammunition to justify their position. The program cancellation would have been disastrous for naval aviation.

Thursday 19 February Ho and I flew 1.8 hours in F2 to the upper altitude limit for the aircraft. Fifty thousand feet is the limit for aircrew to fly without a pressure suit like those worn by the astronauts. The altitude restriction is in place to protect aircrew in the event of a cabin pressurization failure which could cause life-threatening injuries without a pressure suit. We were evaluating the air conditioning system and thrust available at that altitude along with other items. Thrust wasn't a problem as the engines performed well.

Monday 9 March Huffer and I shared a NVG warmup flight on the VR-1754 west of PAX. He was a skilled NVG pilot and it was an enjoyable flight[314].

Saturday 21 March Knockers and I conducted a 2.5-hour wing-drop flight in F2. The strips of wood weren't fixing the problem at this point. Chum and the other SH WSOs started flying the wing-drop flights with the Hi 8mm camera as I stepped back.

We didn't appreciate was how elusive the wing-drop solution would be. It took many months, 10,000 wind-up turns and 100 configurations to finalize a solution (Elward, 2012, pp. 116-121). Ricardo suggested taking off the wing-fold panel as an option. Testing showed that this change eliminated wing-drop but caused new problems. Eventually, a configuration of a porous wing-fold panel allowed high pressure air from under the wing to flow onto the upper wing to eliminate wing-drop. This change was incorporated in production aircraft then

---

[313] https://www.washingtonpost.com/archive/local/1998/02/15/for-navys-new-jet-fighter-a-bit-of-turbulence/0f0c01a9-582b-439c-b30c-dc343e6f76ec/
[314] Huffer ultimately became the Commander, Naval Test Wing at PAX. He and I shared the hobby of model airplanes in our youth; https://www.modelaircraft.org/capt-tom-huffer-huff-usn-commander-naval-test

improved after better solutions were developed over time. What others had tried to use to cancel the program, was not an issue in time.

Wednesday 8 April Chum, Lester and I attended an AESA radar class on base. The system had the potential to revolutionize tactical aircraft employment.

May I started volunteering at the PAX Flying Club. I proposed involvement with the Young Eagles[315] program to inspire youth aged 8-17 with flights in our Cessna aircraft. I also asked the members if they were interested in an aerobatics ride in an open cockpit Pitts S-2A. Our ITT Director, Pete, owned a Pitts and was willing to provide flights. The aircraft was conveniently located at the nearby St Mary's airport. Many aero club members jumped at the idea.

Saturday 16 May We started the Pitts aerobatics demonstrations at 9am and Pete flew minus an occasional refuel until noon. I coordinated the swap of passengers between flights while the Pitts engine was running at idle. Each pilot returned impressed by their experience. I climbed into the forward cockpit as the last passenger of the day. Pete was generous to demonstrate aileron rolls, snap rolls, a split S, a Hammerhead stall and an inverted spin. I can only imagine how Pete felt after performing aerobatics all morning.

Early the next morning, life changed for Wendy and I as our precious daughter, Elizabeth, was born. I took annual leave to help with our new family member.

---

[315] The program inspires youth with flying; https://www.eaa.org/eaa/youth/free-ye-flights.

Friday 5 June The astronaut selection list came out with Strike squadron mates Ken[316], Zambo[317], Dex[318] and Lester[319] making the cut. That was great news! I'd flown with all but Lester (he was an NFO) and they were excellent pilots.

An updated F/A-18F ORD established requirements for a new acquisition contract and specification. The Naval Aviation Requirements branch at the Pentagon led the effort and eventually directed NAVAIR to award a contract to acquire the updated F aircraft. The new ORD was in work as we flew the first two F aircraft.

The WSO in the back seat of an F needed access to superior technology to pursue his/her targets. Plans were in place to purchase the ATFLIR and the AESA radar for the SH. Those two sensors and MIDS could dramatically enhance the warfighting ability of the aircraft. However, the display in the back of the F model did not have sufficient surface area or resolution to take advantage of the improved sensors. The F-15E Strike Eagle rear cockpit had a much larger screen with better resolution to display its targeting information leading to better weapons employment. We recommended a bigger screen with higher resolution combined with a better viewing angle for the WSO.

As the legacy Hornet was primarily a single seat tactical aircraft, the avionics functionality didn't fully take advantage of a WSO in the D or F aircraft. For example, the aircraft mission computer had A/A modes and A/G modes but only one could be selected at a time. Tactically, it is advantageous for the pilot to be defending the aircraft in the A/A mode while the WSO performed targeting in the A/G mode in preparation for a bomb release. This was corrected in a later Block II upgrade with the AESA radar and software that enabled simultaneous A/A and A/G computer modes.

---

[316] Ken as a NASA astronaut; https://www.nasa.gov/sites/default/files/atoms/files/ham_kenneth.pdf. Interesting to see later that he was stationed at NASA after graduation from the Naval Academy.
[317] Zambo as a NASA astronaut. https://www.nasa.gov/sites/default/files/atoms/files/zamka_george.pdf
[318] Dex was an exceptional man who we miss; https://www.nasa.gov/wp-content/uploads/2016/01/poindexter_alan.pdf
[319] Lester as a NASA astronaut; https://www.nasa.gov/sites/default/files/atoms/files/woodward_neil.pdf

Similarly, there are many tactical scenarios when it was advantageous to have the pilot speaking on one radio while the WSO was speaking on a different radio. This ability was common in many older aircraft including the Intruder. However, the Hornet radio design did not support simultaneous radio transmit. A fix was later incorporated in an upgrade.

There were scenarios such as the FAC (A) mission where a pilot needed the moving map to support aircraft navigation, but the WSO needed to access the map in a different area or scale to assist in target identification and weapon designation. Unfortunately, the two cockpits displayed the identical map as they were synchronized. This was later corrected in a Block upgrade with a digital map that allows independent access from either cockpit.

While using the ATFLIR, a WSO needed to control the LASER for precise ranging to targets for weapons release and when designating a target for LASER guided munitions. Unfortunately, the control switch for the LASER resided in the forward cockpit only. A LASER control was later added to the aft cockpit.

The rear cockpit UFCD was difficult to use tactically by the WSOs as it was located low in the cockpit. We preferred the UFCD above the glare shield and behind the pilot's ejection seat. This location could enhance the WSO's situational awareness by displaying the ALR-67 threat radar warning display high in the field of view.

With the relocation of the UFCD, the central grab bar with chaff and flare dispensing buttons on top of the glareshield required relocation adjacent to the UFCD. This grab bar location was preferred as it was helpful for a WSO to cross their body with an arm and pull to twist their torso around for better visual acquisition of enemy fighters or missiles.

As ideas for a F rear cockpit redesign developed, we needed an advocate who could review them and provide a feasibility assessment. Many single seat pilots did not appreciate how the changes could improve their WSO's performance and ultimately make the aircraft a better warfighting machine. They would rather spend money elsewhere.

MCAIR, now Boeing, had developed the two seat F-15E Strike Eagle and knew how to design rear cockpits of tactical aircraft. Lyle and I discussed the changes. He was a master at crew station layout and intimately knew the effort involved in ECP paperwork and political navigation to get the changes approved. Lyle took our recommendations to Boeing management. They were receptive to our ideas and spent their own time and money investigating the feasibility of the changes.

Wednesday 1 July I sent an email to CDR Scott 'Notso' Swift who worked at N88, Naval Aviation Requirements at the Pentagon. The email provided my perspective on improvements to the Super Hornet to enhance its warfighting capability. I focused on acquiring improved resolution for the air-to-ground targeting FLIR and radar, the AESA radar, bigger, high-resolution displays along with the JHMCS in both cockpits, MIDS, PIDS and IDECM. Notso forwarded my email to the USN/USMC Hornet community, complimenting my suggestions and encouraged others to share their thoughts on the SH improvement plan.

Tuesday 21 July Hornet test pilot, callsign Pink and I did a shake test on the new EGI system by completing catapult launches and arrested landings at PAX. The system performed flawlessly despite all the vibration and flight loads. The EGI system added GPS to the Hornet, improved navigation, weapon delivery accuracy and supported JDAM bombs. Its implementation revolutionized the Hornet once again.

Friday 24 July Tomcat test pilot, callsign Rowdy and I had a F/A-18B video chase flight in support of E-4 high AOA testing. I briefed him like the many other pilots I'd flown with on E-4 chase flights. I explained the importance of minimizing accelerations during the transition into the dive to help keep the camera steady. Rowdy acknowledged my concerns and promised to do his best. Once out in the area, we orbited around E-4 as it slowed to stalling speed and departed controlled flight. I was looking through the camera viewfinder intent on filming E-4 as Rowdy maneuvered into the descending spiral. The transition was so smooth than I barely noticed it. The video footage obtained was excellent.

I complimented Rowdy on the entry and asked him to repeat it for subsequent test points. He appreciated the comment and ventured to duplicate the effort. We set up for the next departure and Rowdy flew it like the first pass. I was impressed. I'm not sure exactly what Rowdy did to control the aircraft but he was by far the smoothest of all my video chase pilots. It was as if he had magic hands that produced no bumps or accelerations. He made getting video an easier task enabling me to use a narrower camera field of view. That resulted in better video detail. Despite flying with many other pilots chasing E-4, none matched Rowdy's skill on this task. He later flew the Boeing X-32 aircraft in flight test[320].

Friday 4 September CNO Johnson released a message with Navy-wide distribution encouraging personnel to seek and accept overseas assignments. He highlighted the need to fill overseas billets, the financial and cultural advantages of overseas tours and how they provide once-in-a-lifetime experiences. The message resonated with me after enjoying the tour with the Eagles aboard Midway.

The commander selection board results for my year group were released and I wasn't selected for promotion. That was a significant disappointment. If I failed to promote to commander the following year, a Navy Continuation Board could extend my career to 20 years of active-duty service then my retirement was mandated. Unfortunately, few officers are selected for promotion once 'passed over' on their first attempt.

As my three-year tour at the Super Hornet ITT neared completion, I needed a last set of orders before my retirement at 20 years of active service. I searched the available jobs for AEDO's with rotation dates near the start of calendar year 1999. The search brought up a position in Japan as the Officer in Charge (OIC) of the structural maintenance team at the Naval Air Pacific Repair Activity (NAPRA) Detachment Okinawa (NDO).

My detailer was pleased to hear of my interest in NDO as it wasn't easy to fill overseas billets. In discussions with the detailer, I requested and was approved for orders to C-12 aircraft training prior to moving to Okinawa. This enabled me to serve as a co-pilot on logistics flights throughout WESTPAC.

---

[320] Rowdy Yates, another great test pilot; https://www.pbs.org/wgbh/nova/article/meet-a-test-pilot/

Pistol and I had an afternoon training flight in a F-14. As the sky was a clear and bright blue, he suggested we fly VFR to Manhattan and back. I enthusiastically agreed but noted that I'd never flown VFR through the very busy, east coast airspace near New York. To ensure we didn't create a problem, we checked in and maintained communication throughout the flight with ATC. We flew at 11,500 feet as we proceeded to the northeast from PAX. The ATC controllers were kind and accommodating on our way north. A couple of them asked about our intentions. I explained that we were flying to the Manhattan area and then would be returning to PAX.

As we approached Manhattan, a great feeling of pride welled up inside of us as we looked down on all the great buildings including the World Trade Center. We were flying one of America's premier fighters and displaying it on a beautiful day above New York. We felt proud and patriotic representing our nation in that aircraft. As we got closer to Manhattan, a quickly speaking ATC controller with a thick New York accent called, "Strike 411, what are your intentions?" I replied, "We plan to overfly Manhattan and return to Navy Patuxent River". He sharply replied, "No, that is not going to happen. I've got too much traffic for that. I need you to turn left now to a heading of 190 degrees and return south". "Left to 190 degrees, Strike 411" was my quick reply. We turned and headed back towards PAX retaining a great feeling after seeing Manhattan and the beautiful skyline.

16 September TPS hosted an ALQ-99 Jammer Pod brief in the morning. The system flew on the pylons of the EA-6B. As the F/A-18G Growler was being discussed, using the tactically effective, mature ALQ-99 pods on the SH seemed like a good idea.

September Knockers accepted orders to be a Professor at the USNA[321]. Shortly after his arrival, he wrote an article for the magazine, Wings of Gold, explaining the success and challenges of the SH program[322].

---

[321] Knockers at the USNA; https://www.baltimoresun.com/news/bs-xpm-1998-09-06-1998249024-story.html
[322] Knockers' SH article; https://fas.org/man/dod-101/sys/ac/docs/990414-ART-Super-Hornet.htm

Friday 13 November The Hornet Executive Steering Committee (HESC) released a message outlining their priorities for future development and funding. The HESC was composed of leaders from the Navy and Marine Corps Hornet community. The number one priority was properly funding past shortfalls to keep the fleet aircraft flying. The number two priority was the ATFLIR. Further down the list were GPS, MIDS, PID/CIT, AESA, JHMCS and Expand 4/5 of the radar. The message provided validation for many of the efforts that were ongoing at NAVAIR.

Thursday 19 November Hornet test pilot Don and I shared my last flight in a Navy jet as we ferried a F/A-18B from PAX to NAS NI, San Diego for periodic depot level maintenance. We logged 6.8 hours due to flying against the wind and having to stop twice for fuel. Don was happy for me to manually fly the jet across the country instead of using the autopilot.

A plan was drawn up by the Hornet program office to address the upgraded F/A-18F ORD and WSO concerns. A group of proponents was assembled in 1999 (after my transfer from the ITT) to review the items needed to improve the F warfighting capability. The review was called a Quality Function Deployment (QFD)[323]. The items identified as the most promising were incorporated into the Block II upgrade for the aircraft. Much to our delight, the QFD and Boeing instituted many of the original ITT WSO recommendations into the Block II upgrade of the F/A-18F aircraft. The Block III version rolling off the production line delivers an even greater warfighting capability and reflects additional Strike WSO inputs from the original flight test years.

Late March 1999 The developmental testing of the SH concluded turning the aircraft over for OT[324].

---

[323] QFD is a process and set of tools used to effectively define customer requirements, convert them into detailed engineering specifications and plans that produce products that fulfill the requirements.

[324] The Super Hornet developmental flight test program concluded in April 1999 with over 3100 flights and 4600 flight hours. https://boeing.mediaroom.com/1999-05-05-Boeing-Super-Hornet-Successfully-Completes-EMD

30 March 1999 The Super Hornet OPEVAL commenced at VX-9 at China Lake and concluded in November.

14 February 2000 COMOPTEVFOR released the OPEVAL results citing the Super Hornet aircraft as operationally effective and suitable. A phenomenal amount of work had gone into that positive outcome! This OPEVAL led to a positive production decision and funding for the fleet aircraft purchases.

15 June 2000 Boeing was awarded a $8.9B contract for the delivery of 222 Super Hornets over a five-year period ($40M per aircraft). The improved F/A-18E/F Block III aircraft cost $51M in 2020[325]. F-35 cost estimates in 2021 are $117M per aircraft.[326]

The Super Hornet is an excellent aircraft that is well-suited for carrier aviation. The maximum carrier takeoff and bring-back weights deliver effective warfighting capability and flexibility. The speed, agility, onboard sensors, network intelligence and eleven weapon stations make the SH a potent foe. Improvements to the aircraft since initial flight test have enhanced its capability. To help pilots land aboard the carrier, NAVAIR developed new aircraft software appropriately named 'MAGIC CARPET'[327]. It makes landing aboard the carrier a simpler task and eliminates the past need for extensive shore-based landing practice. While it was a large team effort, manager Buddy Denham, my NAVAIR mate, Steve and Decoy[328] were pivotal in getting Magic Carpet to the

---

[325] Super Hornet Block III aircraft are the latest version; https://www.forbes.com/sites/erictegler/2020/06/22/us-navy-just-got-its-first-new-fa-18-super-hornets---here-are-the-key-upgrades/?sh=7105061b3d38

[326] Each aircraft has advantages over the other; https://armscontrolcenter.org/f-35-joint-strike-fighter-costs-challenges/

[327] MAGIC CARPET is a game changer for naval aviation; https://www.youtube.com/watch?v=FMTf_Z9rMh0&t=34s. As the Navy loves acronyms, here is the ultimate; Maritime Augmented Guidance with Integrated Controls for Carrier Approach and Recovery Precision Enabling Technologies (MAGIC CARPET).

[328] Admiral 'Decoy' Dunaway had an exceptional career retiring after 33 years of active duty. He led NAVAIRSYSCOM during his last tour of duty; https://www.navy.mil/Press-Office/News-Stories/Article/2263969/innovation-marks-retiring-navair-commanders-legacy/

fleet. The SH will serve the fleet crews well for many years to come[329]. As time passed, we gained a better understanding. John Mellencamp's wish had come true for naval aviation.

---

[329] As of April 2020, 322 F/A-18E and 286 F/A-18F aircraft had been manufactured with Block III versions in production. https://www.NAVAIR.navy.mil/news/Navy-takes-delivery-final-Block-II-Super-Hornet-looks-ahead-Block-III/Thu-04232020-1129 and https://www.janes.com/defence-news/news-detail/boeing-delivers-first-production-super-hornet-block-iiis-to-us-navy

# Chapter 16

## Aircraft Structural Repair from Okinawa, Japan

*NDO served as a responsive, mobile Intensive Care Unit (ICU) for aircraft that had structural damage or extensive corrosion prohibiting flight. We strived to have repair personnel onsite within 48 hours of receiving a message from a fleet squadron located anywhere in the Western Pacific[330].*

Friday 4 December I took annual leave, drove down to Richmond International airport and checked into a motel to complete Multi-Engine (ME) training in a PA44-180 Piper Seminole[331]. The training was excellent and the qualification was completed in several days.

Tuesday 15 December Notso provided a letter he wrote supporting my promotion to commander. Notso stunned me with his powerful support and kindness, lifting my spirits and motivating me to continue to give my best[332].

Friday 15 January I assembled a package of information supporting my promotion to commander and provided it to the FY-00 Promotion Board.

Sunday 17 January Scanner (my predecessor at the VX-5 FSO) was the CO at NAPRA in Atsugi. Despite still being attached to Strike and the ITT, he requested my attendance at a series of meetings in Japan. I flew commercially from Washington, D.C. to Tokyo to spend several days in Atsugi. It was great to see Scanner again, meet his XO, Dory, the NAPRA Atsugi staff and Matt, the current NDO OIC.

---

[330] The aircraft were located anywhere from Hawaii to the Persian Gulf and from Northern Japan to Antarctica.

[331] The Seminole was underpowered but good for initial ME training; https://www.piper.com/model/seminole/

[332] This was the last time Notso and I had contact until 2021 when we were both part of a Zoom conference discussing the future of naval aviation. I was unaware of his success. Admiral 'Notso' Swift commanded the SH FRS and retired as a 4-star; https://www.navy.mil/DesktopModules/ArticleCS/Print.aspx?PortalId=1&ModuleId=692&Article=2235990

NAPRA was undergoing a major review to reduce the cost of doing business. The review followed a Business Process Re-engineering (BPR) model[333]. We were examining every part of the organization to find a better way forward. While the meetings at NAPRA were rather intense, it was like going home to visit Atsugi. Many of the Japanese staff at the Officers Club were still present from my days with the Eagles. We were overjoyed to see one another.

Sunday 24 January Once the Atsugi meetings concluded, Matt and I flew two hours south to Naha airport on the island of Okinawa to visit NDO on MCAS Futenma. The tropical climate of Okinawa was warm and welcoming despite it being winter.

Monday 25 January Matt picked me up in the morning from Marine Base Foster. He was a very intelligent, athletic and personable Navy P-3 pilot who loved his tour on Okinawa. We arrived at MCAS Futenma and were sharply saluted by the Marine sentry at the main gate. The NDO staff were very friendly and inviting upon arrival. The detachment had a family-like feel with everyone working together to accomplish their very important task of aircraft structural repair.

We visited the immaculate NDO production hangar, supply storeroom, tool room, machine shop and warehouse. The production supervisor, Duane, had a youthful spirit despite extensive aircraft repair experience.

Matt and I spent a couple days together as he briefed me on the responsibilities of the detachment, introduced me to various partners on the island and shared upcoming tasks that needed attention. It was great working with him as he was clearly motivated to help the fleet.

Thursday 28 January I returned to Southern Maryland from Okinawa. This short trip was the start of many international journeys while working with NAPRA.

Monday 1 February I flew to Wichita, KS to attend C-12 ground school and simulator training at Flight Safety International (FSI). They

---

[333] BPR seeks to optimise a business. https://kissflow.com/workflow/bpm/business-process-reengineering-bpr/

are a premier aviation training company with offices around the world and their instructors were excellent. The Navy C-12 is a Super King Air 200 in the civilian world[334]. I had a simulator session each day from 8-12 February.

Friday 12 February I completed training at FSI and flew to NAS NI to attend flight training with VRC-30[335], the C-12 FRS. The squadron also provides logistics for naval aviation around the world with carrier onboard delivery of personnel, mail, food and other supplies. Our class attended more ground school lessons and took more tests in preparation for flying the C-12. I flew the C-12 from 19 February to 4 March.

I returned to Southern Maryland to take leave, pack our household goods and organize our personal affairs for the flights to Japan. Wendy, Elizabeth and I traveled from Washington, D.C. to LAX then to Yokota AFB near Tokyo then southward to Kadena AFB, Okinawa. Elizabeth was only 10 months old and we were apprehensive about how she would cope with such a long journey. She was a star tolerating all the moving around like a champ.

Saturday 20 March Matt met us at the AMC terminal and drove us to our temporary military family accommodation on Kadena[336]. We were happy to open our suitcases and relax.

Aircraft repairs are generally categorized O-level, I-level or D-level. O-level repairs are managed at the squadron level with their trained sailors. I-level repairs are completed by the Aircraft Intermediate Maintenance Department (AIMD) outside of the squadron yet manned by sailors. These repairs often repair a removeable component from the aircraft. Depot level (D-Level) repairs required specialized tools, materials and uniquely skilled and qualified artisans. D-level repairs were often structural, meaning that a faulty repair could result in the loss of the

---

[334] The Navy flew C-12F aircraft; https://www.airforce-technology.com/projects/c12-huron/
[335] VRC-30; https://www.airpac.navy.mil/Organization/Fleet-Logistics-Support-Squadron-VRC-30/About-Us/History/
[336] A map of Okinawa; https://cnrj.cnic.navy.mil/Installations/CFA-Okinawa/About/Installation-Guide/Installations/

aircraft due to a failure in material or workmanship. Our teams worked diligently to ensure this never occurred.

Major industrial depots were located at various locations around the USA; Norfolk, VA, North Island, CA, Cherry Point, NC and Jacksonville, FL are a few Navy depot locations. The Japanese commercial depot Nippi, located at NAF Atsugi, repaired various Navy aircraft with scheduled, periodic maintenance but didn't respond to emergent aircraft problems. The emergency repairs were handled by NDO.

NDO served as a responsive, mobile Intensive Care Unit (ICU) for aircraft that had structural damage or extensive corrosion prohibiting flight. We strived to have repair personnel onsite within 48 hours of receiving a message from a fleet squadron located anywhere in the Western Pacific[337]. This was not easy to accomplish as repair research was required, parts and tools were collected, travel orders were printed and signed, plane tickets and accommodations were booked. It was a team effort to pull all the pieces together. Once ready, one of the administration staff drove the repair team to the Naha airport.

NDO provided repair support for over 20 different Navy and Marine aircraft types. Each aircraft has its own unique material needs due to their unique design and structural demands. These material demands often required specialized tools to enact the repairs. If a squadron found a crack in a structural frame that was beyond permissible limits, the aircraft was grounded (could not fly) until it could be repaired. The squadron or airwing sent a message describing the downing discrepancy and our detachment jumped into action.

Our engineers and planners researched the area of the damage in technical publications and determined the authorized repairs. If the damage exceeded permissible limits or there weren't any specified repairs, our engineers coordinated with stateside engineers responsible for that aircraft type. A unique repair could be developed and documented with the responsible engineers involved. As the repairs required expertise beyond that available at the squadron level, the work

---

[337] The aircraft could be located anywhere from Hawaii to the Persian Gulf and from Northern Japan to Antarctica.

299

was completed by the NDO staff. If parts were needed for the repair, they were manufactured at NDO by the sheet metal artisans or our machinists who had years of stateside depot experience.

In addition to our quick-response repair efforts, NDO had a Planner/Estimator and two structural repair artisans aboard every aircraft carrier in the Middle East. These carrier teams were replaced every three months with a fresh crew. As the carrier cruises were often six-months long, two teams covered a carrier's deployment. The team carried a laptop full of technical publications along with three large boxes of tools, sheet metal, composite materials and various fasteners. They repaired aircraft that otherwise would have been hangar queens until the completion of the cruise. Communication between our carrier teams and Okinawa was frequently via email when available or naval message. Coordination with the CAG Maintenance Officer (CAGMO) and the admiral's staff could secure more time-critical or classified communication resources.

At times, an aircraft needing repair wasn't aboard the aircraft carrier but was a helicopter aboard one the smaller ships in the carrier battle group[338]. In this case, the repair team were transported to the smaller ship by helicopter to inspect the aircraft and enact the repair. The team carried the tools and materials anticipated for the repair. Occasionally, the repair was ashore requiring a helicopter or COD ride as well.

NDO had a staff of forty-two with twelve being military of various ranks and ratings with the OIC being the only officer[339]. The remainder of the staff were U.S. government and Japanese civilians. The OIC's civilian deputy was a GM-13, an advanced management grade for government employees. Howard occupied the GM position when I arrived but retired to Las Vegas with his wife shortly thereafter. JJ became our new GM after Howard's departure. Ed was the Planning Department supervisor with five planners working for him. Each planner was tasked with inspecting and documenting aircraft damage, researching

---

[338] The carrier battle group (now called carrier strike group); https://science.howstuffworks.com/carrier-group.htm
[339] In 2002, Andy joined NDO. He was a hard-charging ensign that had worked his way up through the ranks.

the technical publications to determine repair protocols, coordinating with our engineers to develop a repair, coordinating with supply to ensure the appropriate tools and materials were available for the repair and finally working with the artisans to ensure they were competent and comfortable with the repair. The planners had a difficult job but each one was hand-picked based on past excellence in their work.

Duncan supervised our sheet metal mechanics, three machinists and general mechanics. Phyllis served as the purchasing officer and managed the supply department comprised of our finance officer, callsign G, along with a few staff members in the warehouse and toolroom. Our two engineers, John and Jim, worked together and shared the load of coordination with stateside engineering competencies as well as design. Our technical publications area had three employees. Several administrative staff were responsible for keeping track of everything including travel orders, tickets for our never-ending revolving door of travelers, reports and outgoing messages. They all did an amazing job. It was a rather small operation, but its importance to naval aviation couldn't be over-stated.

Ed was a joyful character that loved serving naval aviation. He was a retired Navy Senior Chief Petty Officer that wanted to keep supporting the fleet. His dedication was inspiring as Ed's wife and family lived in Norfolk. He lived by himself on Okinawa and returned to the States on leave to see his family when possible. Ed often arrived at work before others despite our working hours starting at 7am. He read the overnight message traffic to determine if we had any urgent tasking and if so, how urgently we needed to respond to the request. He started coordinating immediately and didn't mind handling challenges. Ed's attitude set a standard for the entire detachment. He pursued excellence and encouraged us all to join him in the effort. Ed was a selfless man, sacrificing a great deal to keep the forward deployed warfighters flying.

Wednesday 31 March Matt and I attended the weekly morning meetings at the CFAO headquarters building on Kadena. The meeting provided an opportunity for face-to-face time with naval leadership and for them to quickly distribute critical information to geographically dispersed Navy units onboard Okinawa.

Thursday 1 April Wendy and I found a new, three-bedroom apartment off-base to rent and commenced moving in.

Saturday 10 April I attended my first safety meeting for the Kadena AFB Aero Club (KAC) at the Shilling Community Center. The meeting was well-attended and the briefs were of a professional standard. I looked forward to flying with them[340].

Saturday 29 May Bobby Suell was the long-time manager of the KAC. He had retired from the Air Force as a Master Sergeant after 25 years of service and had been a flight instructor for a long, long time. I completed a Cessna checkout with Bobby.

Thursday 3 June After months of turnover, Matt and I had a Change of Charge ceremony. It was like a COC ceremony but on a smaller scale. We shared the event with the NDO staff, local fleet customers and friends along with Scanner. I spoke of my desire to get back to fleet aviation to serve in a meaningful and helpful manner. We noted that during the recent USS Vinson deployment to the Persian Gulf, NDO teams performed repair on 50% of the carrier's aircraft. Without the NDO team, the carrier's airwing would have been severely crippled and unable to perform their assigned duties.

Despite a concerted effort with supporting documentation, I was passed over again for promotion to commander. I received a letter from the Bureau of Naval Personnel (BUPERS) acknowledging my second failure to promote to commander. I was to be administratively separated from the Navy unless extended on active duty by a Continuation Board.

Wednesday 9 June CFAO held a terrorist awareness briefing that provided an excellent review of the history of events. We were reminded of the attacks on the USMC barracks in Beirut in October 1983, the World Trade Center garage bombing in February 1993, the bombings of US Embassies in Nairobi and Tanzania in August 1998 and the attack on the USS Cole in October 2000. The prevalence for terrorist attacks was increasing. We were all called to be alert for terrorist threats to our personnel and infrastructure.

---

[340] This new video provides a good overview of the Club; https://www.youtube.com/watch?v=YoSW6orcYU4

Thursday 10 June I commenced three weeks of temporary duty in Atsugi as we continued BPR discussions to streamline the NAPRA business. There were major efforts to cut costs and every part of the NAPRA business was on the table. I didn't agree with several proposals as they withdrew support for the forward deployed fleet squadrons.

Sunday 1 Aug Category 1 Typhoon[341] Olga passed over Okinawa with 90mph winds.

Monday 2-4 Aug The results of the NAPRA BPR were briefed to the NDO staff. We lost 5 full-time government positions out of 30 as they were deemed redundant, better managed with headquarters assistance or the existing staff could take on additional workload. A few of the five understood the rationale for the changes and were ready for a move. A couple were very disappointed. I supported greater efficiency if service to the fleet customer didn't suffer in the process. Fortunately, we avoided changes that threatened our superior service.

August Scanner strongly endorsed my officer continuation package citing the critical nature of my work as the OIC at NDO and sustained superior performance. Fortunately, the Board supported my continuation on active duty until twenty years of service. I continued to seek opportunities to strongly support naval aviation in my last few years of duty.

Saturday 11 September Commander Naval Air Force Pacific (COMNAVAIRPAC), Vice Admiral Mike Bowman, visited Okinawa and met Captain Green, the incoming CO at CFAO. Admiral Bowman was our CAG aboard the USS Midway. We'd flown together in an Intruder from the ship, but I wasn't sure if he'd recall as over ten years had passed. I contacted his staff prior to his arrival and suggested a briefing and tour of the NDO repair facilities. Much to my surprise, Admiral Bowman agreed to the idea. On the day of our meeting, I waited on Kadena AFB for the admiral to complete another briefing. We planned for his government, black limousine with blue flags bearing three stars to

---

[341] Hurricanes in the USA are called typhoons in Asia. Category 1 typhoons have winds to 95mph. https://www.nhc.noaa.gov/aboutsshws.php

follow my car to Futenma. I was surprised as the admiral greeted me by my callsign.

When we were ready to depart, I suggested that his transportation follow my car over to Futenma. He responded, "They can follow us. I'm riding with you." That was another big surprise. The trip to Futenma took 20 minutes. We had time to talk about life and family. I mentioned our daughter, Lizzy and a favorite quote of mine from Theodore Roosevelt[342];

'There are many kinds of success in life worth having. It is exceedingly interesting and attractive to be a successful businessman, or railroad man, or farmer, or a successful lawyer or doctor; or a writer, or a President, or a ranchman, or the colonel of a fighting regiment, or to kill grizzly bears and lions. But for unflagging interest and enjoyment, a household of children, if things go reasonably well, certainly makes all other forms of success and achievement lose their importance by comparison.'

Admiral Bowman affirmed the words stating, "I've flown from aircraft carriers all over the world and in the most amazing machines envisioned by man. I've been very fortunate to do just about as much as anyone would ever want to. Yet the most exciting thing in the world for me is to watch my own children succeed. My son is following in my footsteps in naval aviation and witnessing his success brings me more joy than my own."

Upon arrival at the NDO Hanger, VADM Bowman was given a tour of the facilities and manufacturing ability of the team including the machinery, tools, sheet metal manufacturing, composite repair, machining and heat treatment. He recognized that our staff were the depot repair teams onboard the carriers and credited us with keeping the airwings flying while forward deployed.

Tuesday 14 September TPS classmate, Beef relinquished command of his Airwing Five Tomcat squadron, VF-154, at a COC

---

[342] Theodore Roosevelt, An Autobiography by Theodore Roosevelt, 1920, Chapter IX; https://www.gutenberg.org/files/3335/3335-h/3335-h.htm

ceremony in Atsugi. It was great to be present and share the legacy and success of the squadron.

Friday 17 September Tropical Depression 24W formed east of Taiwan heading westerly. We noted its presence but considered it to be a minor threat as it moved northwest toward Taipei. We were wrong.

Monday 20 September 24W slowed its path west, grew intensity into a Category 3 Typhoon Bart and turned 90 degrees to the north tracking directly toward Okinawa. Everyone collected food and water and proceeded to clean areas outside their businesses and homes. Unsecured items could become projectiles in the storm and be lethal. Families congregated at home while television kept us updated on the storm and provided entertainment.

Wednesday 22 September Category 5 Super Typhoon (ST) Bart with winds to 160mph passed 47 miles west of Okinawa[343]. We were astounded at the power of the storm! The wind and rain hit the heavy, metal front door of our apartment with incredible force. For many hours it sounded like a freight train was about to crash through the door. We stood on the back patio (20 yards away from the front door) as it shielded us from the horizontal rain and wind. It was hard to comprehend the force of the storm. Fortunately, the homes on Okinawa are built to endure such storms.

Friday 24 September Bart passed north of Okinawa. Captain Green assumed command of CFAO at a ceremony at White Beach on the eastern side of the island.

October After six months of living in our Japanese apartment, we moved into a single story, typhoon proof, concrete block house on Kadena AFB.

Mid-October My mother passed away in Denver. I booked a flight to the USA and took 10 days of leave to be with family. It was a sad reason to visit Colorado but good to support my father, family and see close friends.

---

[343] 1999 Tropical storm and typhoon history;
https://en.wikipedia.org/wiki/1999_Pacific_typhoon_season

Thursday 4 November CFAO sent off paperwork to COMNAVAIRPAC requesting approval for me to serve as a C-12 copilot.

Monday 15 November COMNAVAIRPAC approved my request to fly in the C-12 aircraft with correspondence signed by Captain 'H-MAC' McDonald, one of my VA-115 Eagle pilots. My name started appearing on the C-12 flight schedule.

Monday 13 December I travelled to NAPRA in Atsugi to attend our COC. It was sad to see Scanner depart but duty called him to his next position.

Tuesday 14 December CDR Freer assumed command of NAPRA with many friends, customers and staff in attendance. I briefed the new XO, CDR Berkin, on NDO business.

Tuesday 28 December After warm-up training flights from Kadena in November and December, I was unleashed for my first cross country as a C-12 copilot. PIC Wayne had prior Hornet time and solid flying skills. He took flying seriously and flew the aircraft by the book. We flew to Osan AFB, South Korea and returned logging 6.8 hours.

February NDO supported aircraft carrier operations, but U.S. based airwing maintenance personnel including the squadron Maintenance Material Control Officers (MMCO)[344] often didn't understand our role or capabilities. Ed encouraged me to travel to the carriers to brief them as they proceeded on cruise to the Arabian Sea and Persian Gulf. We started coordinating with each carrier as it sailed toward WESTPAC. I conducted onboard briefs to introduce our team and explain our role. The brief took about an hour with questions. Initially, the carriers were serving in peace-keeper roles but this changed to combat operations in time. As things developed, we modified the brief to match the circumstances.

---

[344] The MMCO was responsible for the maintenance department's production effort including plans, schedules and supplies.

Monday 7 February My first carrier visit was aboard the USS Stennis while they were at anchor near Hong Kong. They stopped for a port call on their way to the Arabian Gulf. As aircraft carriers traveled to the Middle East, they often stopped in Singapore or Hong Kong.

Tuesday 8 February I waited at the dock for a liberty boat to transport me to the Stennis. There were many supporting personnel waiting to go out to the ship. It was interesting to see U.S. aircraft carriers supported even though they are on the other side of the world. Once aboard the ship's boat, we transited 30 minutes to get to the Stennis. It was anchored well away from the hustle and bustle of the Hong Kong waterways to enhance security.

Once onboard, I met airwing maintenance personnel that led me to a briefing room. I set up a PowerPoint presentation to brief the squadron maintenance teams. My NDO team attended the brief and we had lunch in the wardroom. They went back to work, I departed the ship and returned to Okinawa That was the first of seven trips to aircraft carriers during my time with NDO. Others were not so easy.

Saturday 3 June I traveled to the Air Force Institute of Technology (AFIT)[345], Wright Patterson AFB, Dayton, Ohio for a two-week class on improving Production, Quality and Manufacturing processes. I was impressed by the course reading material, instructors, lectures and facilities.

Saturday 17 June Once the AFIT course completed, I flew to Wichita to attend C-12 refresher training in the simulator at FSI. The training was an excellent review of emergency procedures.

After completing the FSI training, I visited the Cosmosphere Aerospace Museum[346] in Hutchinson, Kansas. The facility has a unique set of displays including a SR-71 Blackbird. They had a space exploration portion of the museum. I descended some stairs and opened the door to see a wall-sized, Nazi flag in front of me. I stood at the door slightly confused then proceeded past it. In front of me looked like a cave, simulating a WWII Nazi Germany V1 and V2 rocket

---

[345] An engineering center of excellence in the USA; https://www.afit.edu/en/
[346] This is a must-see museum for aerospace enthusiasts; https://cosmo.org

manufacturing factory. As I read the plaques, the German investments in aerospace research became clear. They had brought together amazing minds to develop a new generation of technology. I learned of the V3 missile that was designed to be a manned, intercontinental ballistic missile with a pilot on a suicide mission to attack the USA[347]. Fortunately, the V3 rocket was not completed and never used in combat.

Friday 15 Sept The USS Lincoln was scheduled for a port call in Singapore on their way to the Persian Gulf. There was a COD aircraft leaving Okinawa from Kadena AFB to fly aboard the carrier. It was a perfect arrangement and I enjoyed my first COD landing aboard a carrier. We deplaned the aircraft and headed to CAGMO's office. They had scheduled the brief for the evening so I could walk off the ship the next day in Singapore. The brief went well and dinner was a bonus. I slept solo in a 2-man officer stateroom overnight.

At 6am, the Bosun's whistle sounded over the ship's 1MC speaker system. A sailor announced, "Stand by for the admiral". I thought, 'This isn't going to be good. It is Sunday morning. What could be happening on this fine morning requiring the admiral to wake us up so early?' The admiral spoke briefly, "Shipmates, we received orders last night to proceed directly to the Persian Gulf at top speed with no port calls. There will not be anyone allowed off the ship before the Persian Gulf. Communication about our transit is classified and not to be communicated to anyone outside the ship. That is all." Wasn't that just peachy!

Once email was allowed, I wrote to NDO and let them know I'd be delayed but not why. I spent the next 10 days learning aircraft assessment and repair methods from my onboard team, shared meals with them, ran on a treadmill in a gym near the hangar bay and read books. I felt like a stow-away that didn't belong. Others might feel like a prisoner, but I felt fortunate to witness life aboard the USS Lincoln.

As we approached the Persian Gulf, my team provided details about how to get back to Okinawa. They were experienced traveling through Bahrain. I was fortunate to get one of the first CODs off the

---

[347] The evil of 11 September 2001 had been conceived over 50 years before but never occurred.

Lincoln. I returned to Okinawa on Tuesday, 26 September. This USS Lincoln trip was certainly memorable and very different than expected.

Sunday 12 November C12 Pilot Chris and I flew a group of Marines to Guam via Iwo Jima where we stopped for fuel. It was exciting to land on Iwo Jima, the site of a legendary WWII battle between the Japanese, US Naval and Marine Corps forces[348]. It was difficult to imagine the hell that occurred on such a small island in the Pacific as all was peaceful and calm after our landing. We landed in Guam after 6.4 hours of flight time from Kadena. It was beautiful with turquoise water and white sand beaches.

Monday 2 April The morning started like any other, but life was about to change. At 8am, I received a call from a LCDR at Kadena AFB. He asked if I had a STUIII secure phone so that we could discuss a classified matter. I confirmed possession of a STUIII but needed a couple minutes to retrieve the encrypted key to use it. I asked if he could share any details about his call. He declined saying that we would discuss it over the STUIII. That was an unusual response. I inserted my encryption key into the phone and waited for his call.

The phone rang, I answered it and confirmed my readiness to 'go secure'. We pressed our encryption buttons and waited for the link to be established. Once secure, I asked, "OK, what is this about?" The other LCDR responded, "I can't tell you". "Why not? We are on a secure phone", I asked. "It is too classified" he responded. I was confused and stated, "What am I supposed to do? What do you want?" He replied, "Go look at the breaking news on TV and call me back. Here is my number" as he gave me his phone number. That was a crazy phone call. What the heck was this all about?

I found a TV airing news about a navy EP-3 that had made an emergency landing on a Chinese island after a mid-air accident with a Chinese fighter[349]. I returned the call to our EP-3 LCDR and received an

---

[348] President Reagan narrates this short clip about Iwo Jima; https://www.youtube.com/watch?v=vl9mBHe9-kc
[349] The EP-3 event is summarized here; https://www.history.navy.mil/research/archives/Collections/ncdu-det-206/2001/ep-3-collision--crew-detainment-and-homecoming.html

invitation to a meeting aboard Kadena AFB to discuss the matter. They needed our security clearances passed to them before our arrival. We assembled a team to attend the meeting including JJ, Ed, Duane, John and me.

Upon arrival at the EP-3 spaces, our security clearances were verified prior to entry into a secure vault. Military and civilian personnel from all over Okinawa with various expertise were at the meeting. We were told that the aircraft needed a prompt recovery. It was a priceless national asset with highly classified hardware along with a skilled crew. Details of the damage to the EP-3 were shared. We communicated our abilities at NDO to repair the aircraft and explained how it might be possible to fly the aircraft back to Kadena. The room was stunned. They doubted our abilities. We explained who we were, our connections to NAVAIR and the stateside depots, our skilled artisans, experience, parts and tools. We discussed the parts that we might need. Supply personnel provided details of parts in their inventory. The excitement grew with the attendees as we assembled an epic rescue plan.

At NDO, we assembled a team, parts and plans for the recovery. VISA applications were sent to the Chinese Embassy in Tokyo to allow our team to travel to China for the rescue. Our team was ready to travel with parts, tools and personnel in three days. We didn't hear anything from the Chinese Embassy. Initially we thought that their VISA processing was slow. As days turned into weeks, we understood that the delay was intentional. Finally, we heard back from the Embassy. Our team were not allowed to travel to China to work on the plane. After six months, the Chinese government allowed a team from the aircraft manufacturer, Lockheed, to visit China to dissemble the aircraft and fly it away in pieces. In this action, they didn't 'lose face[350]'. Had our team gone to China, repaired the aircraft and flew it home, the Chinese would have been embarrassed. That wasn't going to happen.

In the 6 months prior to the Lockheed team's arrival, the Chinese were able to dissect the aircraft and collect intelligence. It was not surprising that the LCDR from Kadena was cautious to talk to me even over a STUIII. This was a pivotal event in U.S. intelligence security.

---

[350] Saving face is a critical, interpersonal issue in Asia;
https://www.tripsavvy.com/saving-face-and-losing-face-1458303

In early 2001, it became apparent that NDO needed to recruit more talented planners and artisans from the stateside aircraft repair depots. We started communication with the depots around the country to allow us to visit their facilities. Our mission was to draw away their skilled aircraft repair artisans and planners to work for us in WESTPAC. As you might imagine, several of the depots wouldn't even talk to us as they felt our request was beyond rude! Who calls to take away your best and brightest employees? Other depots were polite and hospitable while others fully opened their doors and encouraged our recruitment effort. Many of the depots thought that none of their employees would be interested and we were wasting our time. That turned out not to be true.

Our Production Supervisor, Duane and I left Okinawa on 15 May and returned on 3 June. During this period, we visited the depots at NADEP NI, San Diego, CA, Hill AFB in Ogden, Utah, Tinker AFB in Oklahoma City, Robins AFB, south of Macon, GA, NADEP Jacksonville, FL, NADEP Cherry Point, NC and Scott Air Force Base near St Louis. We gave a PowerPoint presentation that highlighted our facilities and tasking, provided information about housing and pay and took all sorts of questions.

At Hill AFB, we spent time with Sid, a previous NDO employee that was supervising at their depot. He shared this memorable quote; "I miss the aircraft carrier. You can't get that kind of job satisfaction at a big depot. And you can't explain it to folks that have never been there. It is the most exciting place in the world to work". After another briefing, an artisan commented, "You guys have the best mobile tools and materials kits for aircraft repair that I've ever seen". That was great to hear.

As we traveled to various depots, the attendance at our briefs varied. The biggest crowd of interested persons probably numbered twelve. We were about to start our brief at NADEP Cherry Point, NC to two interested employees but that irritated me. I asked the local organizer about others who might be interested. I wondered if people knew we were there giving the brief. I requested and was approved to go into the workspaces looking for others to attend the brief. I was told that a planner, James, might be interested. We found James and invited him to come listen to our brief about NDO. James liked what he heard and applied for a position as a NDO planner. His application was successful and he joined our team on Okinawa.

After 11 September 2001, when American combat aircraft were flying out of Northern Pakistan and needed on-site depot level repair, James was there in combat gear leading a team. God bless you, James. You are a personal hero of mine. You heard the call, answered it and went to a combat zone when folks were in need. You are a great man!

We did attract several other new employees during that recruiting trip. Each one was priceless to us and very much needed as we found ourselves with teams simultaneously on four aircraft carriers after 11 September. We needed the manpower!

Saturday 30 June CDR Berkin relieved CDR Freer as the NAPRA CO.

Tuesday 28 August After work, I flew some Japanese friends on a tour of Okinawa in a C172. Life was about to dramatically change all over the world. We had no perception of the evil ahead.

## Chapter 17

A Family Funeral, 11 September and Aircraft Carriers at War

*The U.S. Navy increased the aircraft carriers in the Arabian Sea to four in response to the Taliban threat in Afghanistan following 11 September.*

Friday 31 August I was scheduled to leave the next day for Singapore to brief the USS Vinson as they headed toward the Indian Ocean. My father called from the United States with tragic news that my greatly loved nephew had unexpectedly died in Australia. Our family was heart-broken and my sister, Janet, needed me in Australia for support. I cancelled the trip to Singapore, took annual leave and bought a commercial ticket to Brisbane departing 1 September.

Wednesday 12 September As I was leaving Australia for my return to Okinawa after the funeral, my niece, Zanna, drove me to the Brisbane airport. As we stood in the line to check my bag, there was a large crowd of people standing around TV screens. I asked Zanna to check out the attraction. It looked like a Hollywood action thriller as we stood a great distance away. She returned to explain that commercial airliners had flown into the Twin Towers in New York City. The world had suddenly changed. I needed to get back to Okinawa ASAP.

I had a short flight to Sydney to catch the long, non-stop flight to Osaka. As we waited for our flight to Sydney, we knew more about the highjacked planes and wondered if more hijackings were in work. I scanned the travelers at our gate to see if anyone looked like someone who would attempt such a horrific act. No one on our flight looked capable of such an atrocity. The flight proceeded without any issues with a somber group of passengers.

Once in Sydney, I walked over to the international terminal and entered the immigration clearance queue. The immigration officer checked my US passport and said, "This is truly a dark day in the history of the world. I am very sorry". Those words of kindness on a tragic day really triggered my emotions. I thanked the agent and moved into the concourse area. As I reached our gate, I again scanned the travelers for

313

anyone that might want to hijack our aircraft. The passengers, staff and crew were edgy as we all knew what had happened in the USA. The airline staff were cautious yet polite.

We were told that due to the events in the United States, the airspace east of Australia was closed. The global skies became void of traffic on 11 September in a very eerie transition[351]. The FAA ATC serviced the Pacific Ocean as far west as Australia and had shut down operations. Our flight path had to be modified as we couldn't fly directly north. We had to fly well inland over Australia and use Australian, Filipino and Japanese controllers to get to Osaka. This ultimately delayed our arrival. Everyone onboard the flight appreciated the sensitive nature of flying on this day and were very considerate of one another.

As we neared Osaka, I noticed several of the flight attendants crying in the galley area. I asked one of the attendants if this was due to the events in the United States. The attendant explained that while we had been airborne, their airline, Ansett, had announced that they were bankrupt and shutting down operations. The flight attendants had just lost their jobs and were uncertain how they might return home. Life was already getting worse for people at great distances from the World Trade Center.

After landing in Osaka, I checked for my connection flight. Okinawa was enduring Category 2 Typhoon Nari[352] with winds to 100mph. There were no flights until possibly late the next day. I secured a hotel room near the airport and joined the world in watching the horrific news of the terrorist attacks. I highly recommend watching 9/11; Cleared For Chaos[353] and the short video of Boatlift 9/11[354]. These videos provide powerful glimpses into the character of Canadians and Americans, respectively.

---

[351] This is an air traffic mapping for 11 September;
https://www.youtube.com/watch?v=xh4V9PZT2VY
[352] Typhoon Nari stalled on Okinawa;
https://en.wikipedia.org/wiki/Typhoon_Nari_(2001)
[353] 9/11; Cleared for Chaos; https://www.youtube.com/watch?v=Zqf1mSuyd9w
[354] Boatlift 9/11 narrated by Tom Hanks;
https://www.youtube.com/watch?v=18lsxFcDrjo

Thursday 13 September I waited around the Osaka airport for many hours as several flights to Okinawa were scheduled then cancelled. Another flight was scheduled for the late afternoon. We boarded the aircraft and took off. We were told the flight conditions were still rough due to the typhoon with considerable turbulence and windshear and we might need to return to Osaka. Fortunately, the winds were gusting right down the runway making the approach to the south slow but relatively controlled. The pilot made rapid throttle changes on the approach for fifteen minutes to counter all sorts of windshear, but we landed safely. I was very impressed with the pilot's efforts!

As we deplaned the aircraft, the airport was a ghost town with no one present. That was mostly due to the typhoon. I took a taxi to the nearby military parking lot to retrieve my car. The lot was locked without an attendant. I rang an emergency number and an attendant allowed me to retrieve the car after a twenty-minute wait. He explained that everyone was in lockdown due to the typhoon. He was surprised to hear that I'd just flown in from Osaka. The primary road to Kadena that was normally filled with traffic was completely empty.

The Kadena front gate was locked upon arrival. That had never happened before. The guards carefully looked at my car sticker and my ID card. They explained that the base was in lockdown for the typhoon and the events back in the USA. I confirmed their words but explained that I'd just flown in from Osaka and lived on Kadena AFB. They searched my car, opened the gates and requested that I drive directly home and quarantine. I was more than willing to comply. Wendy and Elizabeth were happy to have me home. I came across these impressive words from another military member that was on Kadena AFB for 11 September and later served in Afghanistan[355].

Security at all the military bases around Okinawa was very tight after Typhoon Nari cleared. There were long lines at the base entry points. Previously used entry points were locked and barricaded adding to the morning congestion. Vehicle entry to the bases followed a S-pattern around concrete barricades. Each vehicle was stopped and thoroughly searched for weapons or bombs. Security teams at the gates were well

---

[355]https://www.afcent.af.mil/News/Commentaries/Display/Article/223324/remembering-911-commentary/

armed and wearing body armor. It took up to an hour to get through security. Everyone arrived early at the entry points to get to work at a normal time. As others around the world wondered if they were next on the terrorist hit list, we shared their concerns.

Okinawa has a strategic role in the U.S. military presence in Asia. Critical assets are located on the island. As military families on Okinawa, we knew the seriousness of what was happening and prepared as best we could. I began a daily visit to the military communication center to read secret messages about the global terrorist threat with specific concern for Okinawa and our deployed teams. There was a variety of intelligence suggesting other terrorist cells existed in various countries. The US Embassy in Tokyo had issued a warning to US military in Japan on 7 September indicating there was evidence of a credible threat to US forces[356].

I briefed our personnel regarding any substantiated threats to the military bases on Okinawa and our organization. We discussed procedures to mitigate the risks and outlined recommended responses to threats. In many ways, we should have all felt confident regarding our security. The Japanese authorities were meticulous in their management of foreigners visiting from abroad. Okinawa was not a soft target. We had more U.S. military per square mile on Okinawa than anywhere else in the world. Unfortunately, terrorists prefer soft targets and killing defenseless citizens. That is why they are pursued by the peaceful nations of the world.

As the US military response to the attacks developed, NDO was intertwined in the planning and execution. We usually supported one carrier in the Middle East. Our personnel and their families were accustomed to them being gone for three months aboard ship then being on Okinawa for a year. As the carrier force in the Middle East grew to four over the next few months, our personnel stepped up and manned each one simultaneously. We knew that our contributions could be critical to the success of upcoming combat operations as aircraft might suffer considerable damage. We weren't sure how long the teams would be onboard or exactly what they would be doing. As time passed, combat

---

[356] Terrorist warning; https://abcnews.go.com/International/us-warns-terrorist-attack-japan/story?id=80556

operations extended to Northern Pakistan resulting in a grounded aircraft with depot-level damage. NDO deployed a team and repaired the aircraft. I was concerned for the safety of our men but knew that as mission essential personnel, the military would protect them.

Saturday 29 September Flight operations were suspended for weeks as security was reassessed. C-12 pilot Bill and I finally flew passengers from Kadena to Misawa via Iwakuni, landed at Atsugi and returned non-stop to Kadena. Flight time was 11.6 hours.

Tuesday 27 November An NDO planner, two artisans and I landed aboard the USS Stennis off the coast of Taiwan via a COD from Kadena. Stennis was headed to the Arabian Sea to relieve the USS Carl Vinson during Operation Enduring Freedom. After landing and meeting the CAGMO, we briefed the squadron MMCOs on NDO's repair capabilities and answered their questions. The COD flew me to Hong Kong the next morning for a commercial flight back to Okinawa.

Wednesday 9 January 2002 NDO hosted a meeting about the upcoming H-1 Integrated Maintenance Concept (IMC) with the aircraft PMA and MALS-36. Preplanned depot level maintenance was scheduled for the USMC H-1 helicopters in the NDO hangar aboard MCAS Futenma. The work was to be performed by the NIPPI Aircraft Company who had a long-established contract to perform depot-level aircraft maintenance in Atsugi. The meeting generated a Memorandum Of Understanding (MOU) that set the wheels of progress in motion.

Monday 4 March After considerable planning and discussion, H-1 IMC commenced at NDO. The NIPPI team were exceptional in their attention-to-detail. The NDO staff were pleased to be working side-by-side with them. The MALS-36 leadership were very appreciative of the program.

Monday 18 March The U.S. Navy increased the aircraft carriers in the Arabian Sea to four in response to the Taliban threat in Afghanistan following 11 September. The USS Kennedy was normally an east coast ship with deployments to the Mediterranean Sea but had transited the Suez Canal to join combat operations. Due to their past operating region, they were unfamiliar with our capabilities and connections. An NDO brief could provide essential information for Kennedy's cruise.

I flew to Bahrain awaiting further transportation to the Kennedy in the Arabian Sea. A COD flew a few of us, mail and aircraft parts aboard the ship the next day. As I departed the COD, there was something wrong. The flight deck was quiet at 10am. Normally, a carrier was full of noise and activity by this time. There was little happening and few people on the deck.

As we entered the ship passageways, something was clearly different and eerie. There was no-one in the passageways. "Where are all the sailors?", I asked. "They are asleep. We are on Vampire hours" was the response. "Vampire Hours? What are Vampire Hours?", I asked. "We sleep during the day and fly at night. We have breakfast at 6pm, lunch at midnight and dinner at 6am then go to sleep" was the response. I was stunned. I'd never seen anything like it. It made sense as the threats to our aircraft are significantly reduced at night. We fly with lights out to deny visual acquisition in the dark. Enemy fighters are less of a threat as they aren't trained as well as our aircrews nor have the support to be an effective deterrent at night.

As I traveled the ship looking for people, it became obvious that I was wasting my time. The people I needed to talk to were asleep and wouldn't be awake for hours. I waited in the wardroom while trying to learn as much as I could from anyone who was awake. The ship was serving mid-rats[357] at noon.

My contacts aboard ship woke around 5pm and we discussed the briefing. It took a day to organize all the appropriate personnel to attend. As everyone was focused on conducting combat operations, the audience was limited to ten key leaders. They appreciated the brief and knowing our team was aboard the carrier. As anticipated, they weren't familiar with NAPRA, our onboard NDO team or the difference they could make to a cruise.

---

[357] Mid-rats are short for middle of the night rations;
https://www.dvidshub.net/news/255079/midrats-sailors-favorite-meal

After the brief was completed, I was ready to return to Okinawa. The departure wasn't as easy as my arrival. It was like the Hotel California, 'We are programmed to receive. You can check out any time you like, but you can never leave!' I discussed my desire to fly back to Bahrain with the CAGMO. He understood completely but had no flights to offer. He asked me to visit him every day at 7pm to check on the availability of a COD for my return flight. I received a similar response on successive nights, "Looks like we might get a COD in a couple of days".

On the 3rd night, the story changed. CAGMO said, "We have a COD coming aboard tomorrow but it is delivering aircraft parts to an airfield in northern Pakistan on its way back to Bahrain. You can't go". "Why can't I go?" was my response. "Because they are going to Northern Pakistan near the Afghani border and it is a combat zone" replied the CAGMO. "I am combat aircrew. I need to get back to Okinawa. I am okay with going to Northern Pakistan" came my response. CAGMO replied, "I'll see what I can do. Come back at midnight". I went to breakfast and sat on my hands until midnight. I arrived back at CAGMO's office and he said, "Ok, you're manifested on the COD for departure tomorrow at 11am. It is going to Northern Pakistan. Are you sure you want to go?" "Yep. Happy to go. Thank you" was my response.

I was overjoyed to get aboard the COD the next day. I loved being aboard the Kennedy during the middle of combat operations and wished to be serving in combat. However, it wasn't my time and place to stay and other important work needed me back on Okinawa. There were replacement parts on the aircraft with me. The catapult launch felt strong and after a few seconds we were flying away from the sea. We climbed to thirty thousand feet then started cruising above the Pakistani terrain.

As we arrived above our destination airfield, the aircraft slowed down, dropped its flaps and landing gear and started a rather steep but slow, spiraling descent. I'd never seen an approach like this before. The pilot entered downwind and executed a normal landing. I asked the crew about the descent. They explained that only a narrow cone of airspace had been declared safe from shoulder-fired, heat-seeking missiles so they needed to descend in that cone. It took continuous patrols to ensure that airspace remained safe and more patrols to expand the safe airspace. They had limited manpower to clear the area around the airfield.

After landing, we lowered the aft end of the COD to deliver our aircraft parts. The crews were overjoyed to see us. It was warm despite it not yet being summer. The parts were off-loaded, and a couple of maintenance personnel joined me on the COD to travel to Bahrain and eventually back to Kennedy. After takeoff, we climbed with gear and flaps up inside the cone above the airfield until we had reached our cruise altitude. Once at the cruise altitude, we returned south over Pakistan.

We landed in the United Arab Emirates to refuel. Everyone was happy to be back on land and away from the hostilities. Our flight continued uneventfully to Bahrain. I secured a commercial flight back to Okinawa via Singapore, arriving on 24 March. I was very grateful for the journey but was happy to be back home.

Sunday 2 June The NDO staff didn't feel fully supported by the staffs at NADEP NI and AIRPAC in San Diego. I was asked to take my carrier brief to them and explain our role, capabilities and support needed. The meetings were very productive with the staffs becoming much more aware of NDO and how they could support us. The meeting with AIRPAC identified the need for an Intermediate Maintenance Program (IMP) for the new MH-60S helicopters that were due to arrive in Guam in March 2003. I took an action item to investigate our plans[358]. The IMP would be similar the H-1 IMC aboard Futenma.

Tuesday 23 July My June trip to NADEP NI secured an engineer with composite repair expertise to visit NDO for several days to instruct our teams in Super Hornet repair procedures.

Tuesday 3 September My sixth visit to brief the airwing maintenance staff was aboard the USS Lincoln. Unlike my first Lincoln brief, the ship was in port in Singapore. I walked aboard and walked off without any problems, CODS or concern for safety in a combat zone. It all seemed anti-climactic after all the other challenges of getting on and

---

[358] I discussed the MH-60S IMP with the NAPRA's CDR Borno. He explained the challenges of the program due to the aircraft design and historical issues with helicopter depot maintenance. I became the project lead.

off carriers. The airwing brief was a success and our NDO team settled in for their cruise.

It was great to see VFA-115 Eagle Super Hornets aboard the ship on their first operational deployment. I made it to the Eagle Ready Room to visit my Vampire squadron mate, Zoil as he was commanding the squadron[359]. Unfortunately, he didn't know about my trip and wasn't aboard the ship[360].

Saturday 5 October The NDO OIC lineage had been Ed, Matt, myself and now Ed was back to replace me. He and his family arrived on Okinawa and settled back into life as I put the finishing touches on a turnover folder for him.

Thursday 7 November Ed relieved me as the NDO OIC with the NAPRA CO supervising the event. I stayed onboard NDO assisting Ed with special projects including the HH-60S helicopter IMP for HC-5[361] aboard NAS Guam.

Monday 18 November I caught a COD from Kadena and flew aboard the USS Constellation to brief the maintenance teams as they transited to Hong Kong. The brief was well received by the maintenance staff. The COD returned me to Kadena on November 20th making it my 7th, last and easiest visit to an aircraft carrier.

Friday 3 January 2003 I took Christmas leave in Australia and returned to Okinawa to complete my last few months of active-duty service. Wendy and Elizabeth stayed in Australia to enable her start of kindergarten.

---

[359] Zoil enjoyed a successful naval career. He led the Super Hornet OPEVAL, flew the aircraft in combat over Iraq and Afghanistan while as the CO of the Eagles and served as a Hornet Deputy Program Manager. Zoil had acquisition roles before assuming command of COMOPTEVFOR, supervising all Navy operational testing. https://www.navy.mil/DesktopModules/ArticleCS/Print.aspx?PortalId=1&ModuleId=692&Article=2236084
[360] Super Hornet aircraft from VFA-115 logged their first combat missions six days later, on 11 September 2002.
[361] HC-5 is now HC-25; https://www.seaforces.org/usnair/HC/Helicopter-Combat-Support-Squadron-5.htm

I served as a KAC flight instructor and decided to pursue an Airline Transport Pilot (ATP) license. I found a company in the USA that combined the ATP with Boeing 737-NG training. They mailed me ATP and 737 manuals to study for a course commencement in April.

Tuesday 7 January I attended a NAPRA COC in Atsugi with CDR Dorn relieving CDR Berkin with CDR Borno taking over as the XO.

Ed had me prepare a report outlining the infrastructure required to establish a HH-60S helicopter depot level repair capability aboard NAS Guam. In talking with fleet 60S maintenance personnel, they were concerned about the navy taking the inexpensive option for IMP resulting in greater costs down the line. Those words were powerfully presented and shared up the chain in my final report. I tried to protect HC-5 from a painful maintenance decision. I left Okinawa before the decision was made and a contract was awarded.

Saturday 15 March New CFAO pilot, LJ[362] conducted my annual instrument check in the C-12 aircraft during a return flight to Osan AFB. I didn't know it at the time, but this was my last flight in the Navy. Preparations for my retirement and the move from Okinawa got in the way and flying the C-12 became less of a priority.

Sunday 30 March My sister, Joan, her husband, Phil and my niece, Barbara arrived from Colorado for my retirement ceremony. I was honored to have them present.

Tuesday 1 April My retirement ceremony was held in the NDO hangar with various staffs, family and friends in attendance. The IIIrd Marine Expeditionary Force band superbly played music including the National Anthem. The NAPRA CO, CDR Dorn, presided over the ceremony while Ed served as the Master of Ceremonies. CFAO, Captain Kluckman, provided kind words as my guest speaker. Chaplain John gave the invocation at the start of the ceremony.

---

[362] LJ overflowed with joy and enthusiasm for flying.

In my retirement speech, I thanked the NDO and CFAO staffs for their support in my last few years. I thanked the fleet maintenance crews, the C-12 crews at CFAO and maintenance personnel at the KAC. The end of the speech follows; "Peace is our mission. Peace can be achieved in many ways with different people or cultures. One thing history has taught us is that the longest lasting peace is made when you can convince your adversary it is not in their interest to press the option of war. Peace is achieved through strength. And strength is what the U.S. military represents. Strength is what this group today represents. I am honored to have served and to call you friends. God bless you all in all your continued work".

Wednesday 2 April Despite having my retirement ceremony, I was not yet retired. The ceremony provided an opportunity for everyone to come together and share the moment. I had more work to complete for Ed and administrative chores to complete before leaving Okinawa.

Before leaving the island, I took a day of leave and went on a battle-site tour with Marine MWR. We visited a cliff where the tour guide shared the heroic story of the Medal of Honor recipient, Corpsman Desmond Doss[363]. After hearing the story and seeing the cliff, I hoped that someone, someday would recognize the miracles that occurred there and make a movie. Many thanks to Mel Gibson and the entire crew of the movie, Hacksaw Ridge[364].

Tuesday 8 April I flew from Okinawa to Seattle to commence pre-separation processing at Naval Base Bangor, Washington. This is standard procedures for all naval personnel that retire from a WESTPAC duty station. My ATP and 737 studies continued.

While at Bangor, I received word that the previous manager of the KAC, Bobbie Suell had died in an aircraft accident while serving as a flight instructor on 4 April [365]. Bobby was a confident yet quiet aviator

---

[363] Desmond Doss is worthy of our greatest admiration and respect; https://desmonddoss.com/bio/bio-real.php
[364] The movie Hacksaw Ridge captures a horrific WWII battle on Okinawa and the story of an unlikely hero; https://www.youtube.com/watch?v=s2-1hz1juBI
[365] Article regarding Bobbie; https://lasvegassun.com/news/2003/apr/07/certification-flight-ends-in-tragedy-for-student-i/

with over 15,000 flight hours. Bobbie's passing was a terrible loss for all of us that knew him.

Friday 18 April I finally completed naval administrative separation paperwork and started terminal leave[366]. I took public transportation to SEATAC airport and commenced the ATP / 737 training course. I flew four full-motion and visual simulator sessions before an oral exam and successful check ride with a FAA examiner in the simulator. The skill, professionalism and excellence of all the pilots I'd flown with over many years enabled my success. Watching them fly, then flying various jet aircraft from the back seat under their supervision provided the skills needed for the check-ride. Credit goes to each of them for their example as well as trust and confidence in me.

---

[366] Terminal leave was annual leave that was taken at the end of active duty.

# Chapter 18

Retirement

My journey with the United States military provided priceless experiences founded on excellence, dedication, professionalism and a genuine desire to maintain or return peace in a troubled world. My goal in this memoir was to provide a portal for the reader to better see and appreciate naval aviation excellence. Every day there are aviators flying dangerous missions and landing aircraft aboard ships far from home in the middle of the night. The legacy of excellence continues despite it being hidden from our conscious view.

I hope this book might help people understand the mindset and motivation of late 20th century naval aviators. It might dismiss many myths about the United States, the goals and behaviors of the military and might provide insight into the heart of Americans. As the world continues to strive for peace in all corners of the globe, I would hope that nations could see the United States as a friend of peace.

I have felt God's encouragement and peace when doing things beyond my comfort level – including writing this book. I have felt His righteousness when persevering through adversity. I have felt His guidance while participating in exercises with former citizens of East Germany or interactions with Alex Zuyev. I continue to seek His purpose for my life.

I have been fortunate to travel to many countries around the world and witness many cultures cherishing the task of raising their children in peace. We pray for this basic right for those in conflict. Raising our children in peace to grow stronger and better than us should be a fundamental right for all of humanity.

We would hope that in years to come, international terrorism and war are topics for history students rather than a plague on humanity. As we see ongoing battles in Israel, Iraq, Afghanistan, Syria, Sudan, Ukraine and others, we are saddened that the violence continues. We all hope that future generations gain a better understanding.

# References

Springsteen, B. & Bailey, E. (1998). The F/A-18E/F: An Integrated Product Team Case Study. Institute for Defense Analysis.

Elward, B. (2012). The Boeing F/A-18E/F Super Hornet & EA-18G Growler: A Developmental and Operational History. Schiffer.

Lacy, J. (1992). Systems Engineering Management, Achieving Total Quality. McGraw Hill

Pyle, R. (1991). Schwarzkoff; The Man, the Mission, the Triumph. Mandarin.

Powell, C. (May 1990). Proceedings, U.S. Naval Institute Press.

Zuyev, A. & McConnell, M. (1992). Fulcrum; A Top Gun Pilot's Escape from the Soviet Empire. Warner Books.

## Acronyms

| | |
|---|---|
| 1MC | Loudspeaker system used aboard ship to communicate to the crew |
| A/A | Air-To-Air |
| A/G | Air-To-Ground |
| AAA | Anti-Aircraft Artillery |
| ACLS | Automatic Carrier Landing System |
| ACOTD | Assistant Chief Operational Test Director |
| ADI | Attitude Direction Indicator; primary display for the A-6E pilot |
| AEDO | Aerospace Engineering Duty Officer |
| AESA | Advanced Electronically Scanned Array |
| AFB | Air Force Base |
| AGL | Above Ground Level |
| AIMD | Aircraft Intermediate Maintenance Department |
| ALR-67 | Upgraded Radar Warning Receiver |
| AMRAAM | Advanced Medium Range Air-to-Air Missile |
| ANOVA | Analysis of Variance |
| AOM | All Officers Meeting |
| APML | Assistant Program Manager Logistics |
| ASAP | As Soon As Possible |
| ASPJ | Airborne Self Protection Jammer |
| ASR | Approach Surveillance Radar |
| ASSTA | Avionics Software System Test Aircraft |
| ATARS | Advanced Tactical Airborne Reconnaissance System |
| ATC | Air Traffic Control |
| ATFLIR | Advanced Targeting FLIR |
| ATIS | Automatic Terminal Information Service |
| ATP | Airline Transport Pilot |
| BLIN | Built In Test Logic Inspection |
| BN | Bombardier Navigator |
| BOQ | Bachelor Officers Quarters |
| BPR | Business Process Re-engineering |
| BUPERS | Bureau of Naval Personnel |
| CAG | Carrier Air Group Commander |
| CAGMO | CAG Maintenance Officer |
| CAP | Combat Air Patrol |

| | |
|---|---|
| CAS | Close Air Support |
| CATCC | Carrier Air Traffic Control Center |
| CCIP | Continuously Calculated Impact Point |
| CDD | Capability Development Document |
| CDR | Commander; Critical Design Review |
| CFAO | Commander Fleet Activities Okinawa |
| CIA | Central Intelligence Agency |
| Class Desk | The officer in charge of engineering for an aircraft at NAVAIR |
| CMO | Coffee Mess Officer |
| CNFJ | Commander Naval Forces Japan |
| CNO | Chief of Naval Operations |
| CO | Commanding Officer |
| COC | Change of Command |
| COD | Carrier Onboard Delivery |
| COEA | Cost and Operational Effectiveness Analysis |
| COMNAVAIRPAC | Commander Naval Air Force Pacific |
| COMOPTEVFOR | Commander Operational Test Force |
| COTD | Chief Operational Test Director |
| COTS | Commercial Off The Shelf |
| CTP | Chief Test Pilot |
| CU | University of Colorado |
| CVIC | Carrier Intelligence Center |
| CVRS | Cockpit Video Recording System |
| D-Level | Depot level |
| DACM | Defensive Air Combat Maneuvering |
| DFCS | Digital Flight Control System |
| DFIRS | Deployable Flight Incident Recorder Set |
| DH | Decision Height; minimum altitude for a precision approach |
| DOD | Department of Defense |
| DMZ | Demilitarized Zone; buffer area between South and North Korea |
| DPRO | Defense Plant Representative Office |
| DRB | Deficiency Review Board |
| DRS | Detecting and Ranging Set |
| ECMO | Electronic Counter Measures Officer |
| ECS | Environmental Control System |

| | |
|---|---|
| EGI | Embedded Global Positioning System Inertial Navigation System |
| EGT | Exhaust Gas Temperature |
| EMCON | Electronic Emissions Control |
| EP | Emergency Procedure |
| ERS | Emergency Recovery System |
| ESM | Electronic Surveillance Measures |
| EW | Electronic Warfare |
| FAA | Federal Aviation Administration |
| FAC | Forward Air Controller |
| FCF | Functional Check Flight |
| FL | Flight Level: MSL altitude to fly with 2992 altimeter setting |
| FLIR | Forward Looking Infrared Receiver |
| FMS | Flight Management System |
| FPM | Feet Per Minute |
| FOV | Field Of View |
| FRBN | Fleet Replacement Bombardier Navigator |
| FRP | Fleet Replacement Pilot |
| FRS | Fleet Replacement Squadron |
| FSI | Flight Safety International |
| FSO | Future Systems Office |
| FTE | Flight Test Engineer |
| GCA | Ground Controlled Approach |
| GD | General Dynamics |
| GE | General Electric |
| GFE | Government Furnished Equipment |
| GFR | Government Flight Representative |
| GG | Gabrielle Giffords |
| GPS | Global Positioning System |
| GQ | General Quarters |
| GS | Ground Speed |
| GSE | Ground Support Equipment |
| HARM | High Speed Anti-Radiation Missile |
| HESC | Hornet Executive Steering Committee |
| HMD | Helmet Mounted Display |
| HUD | Heads Up Display |
| ICS | Intercom Communication System |
| ICU | Intensive Care Unit |

| | |
|---|---|
| IFF | Identification Friend or Foe |
| IFR | Instrument Flight Rules |
| IMC | Instrument Meteorological Conditions |
| | Integrated Maintenance Concept |
| IMU | Inertial Measuring Unit |
| IMP | Intermediate Maintenance Program |
| INS | Inertial Navigation System |
| ITT | Integrated Test Team |
| IUT | Instructor Under Training |
| JDAM | Joint Direct Attack Munition |
| JFK | John Fitzgerald Kennedy International Airport |
| JHAPL | John Hopkins Applied Physics Laboratory |
| JHMCS | Joint Helmet Mounted Cuing System |
| JO | Junior Officer; generally considered below O-4 |
| JSC | Johnson Space Center |
| JSIPS | Joint Services Imagery Processing System |
| KAC | Kadena AFB Aero Club |
| KGS | Knots Ground Speed |
| KH | USS Kitty Hawk; CV-63 |
| KIAS | Knots Indicated Airspeed |
| KPP | Key Performance Parameters |
| KTAS | Knots True Air Speed |
| Knots | Nautical miles per hour |
| LA | Los Angeles |
| LASER | Light Amplification by the Stimulated Emission of Radiation |
| LAX | Los Angeles International Airport |
| LCDR | Lieutenant Commander |
| LGB | LASER Guided Bomb |
| LT | Lieutenant |
| M | Mach number |
| MAGIC CARPET | Maritime Augmented Guidance with Integrated Controls for Carrier Approach and Recovery Precision Enabling Technologies |
| MAWS | Medium Attack Weapons School |
| MB | Martin Baker Corporation |
| MCAIR | McDonnell Douglas Aircraft |
| MCAS | Marine Corps Air Station |
| MIDS | Multifunctional Information Distribution System |

| | |
|---|---|
| MITP | Missile Integration Test Plan |
| MMCO | Maintenance Material Control Officer |
| MNS | Mission Need Statement |
| mph | miles per hour |
| MSL | Mean Sea Level |
| MWR | Morale Welfare and Recreation |
| MO | Maintenance Officer |
| MOU | Memorandum Of Understanding |
| NA | Night Attack |
| NADEP | Naval Aviation Depot |
| NAF | Naval Air Facility |
| NAPRA | Naval Air Pacific Repair Activity |
| NAS | Naval Air Station |
| NASA | National Aeronautics and Space Administration |
| NATOPS | Naval Aviation Training and Operating Procedures Standardization |
| NAVAIRSYSCOM | Naval Air Systems Command (NAVAIR for short) |
| NAWC | Naval Air Warfare Center |
| NDB | Non-Directional Beacon |
| NDO | NAPRA Detachment Okinawa |
| NFO | Naval Flight Officer |
| nmi | Nautical Miles |
| NPGS | Navy Post Graduate School |
| NROTC | Naval Reserve Officer Training Corps |
| NVG | Night Vision Goggle |
| NWC | Naval Weapons Center |
| OAG | Operational Advisory Group |
| OBOGS | On-Board Oxygen Generation System |
| OCS | Officer Candidate School |
| ODO | Operations Duty Officer |
| OFT | Operational Flight Trainer |
| OIC | Officer In Charge |
| ONI | Office of Naval Intelligence |
| OOC | Out Of Control |
| OPEVAL | Final Operational Evaluation |
| OPSO | Operations Officer |
| ORD | Operational Requirements Document |
| OT | Operational Test |
| OTC | Operational Test Coordinator |

| | |
|---|---|
| OTD | Operational Test Director |
| OTG | Operational Tactics Guide |
| PALS | Precision Approach Landing System |
| PAR | Precision Approach Radar |
| PAX | NAS Patuxent River, MD |
| PDR | Preliminary Design Review |
| PEO | Program Executive Officer |
| PI | Philippines nickname |
| PIC | Pilot In Command |
| PIDS | Positive Identification System |
| PMA | Program Manager |
| PMR | Program Management Review |
| PSD | Personnel Support Detachment |
| PTT | Part Task Trainers |
| QE | Qualitative Evaluation |
| QFD | Quality Function Deployment |
| RAAF | Royal Australian Air Force |
| RAN | Royal Australian Navy |
| RBS | Radar Bomb Scoring |
| RFP | Request For Proposal |
| RIO | Radar Intercept Officer |
| RPM | Revolutions Per Minute |
| ROD | Rate of Descent |
| RUG | APG-73 Radar Upgrade |
| SAR | Synthetic Aperture Array |
| SATCOM | Satellite Communication |
| SETP | Software Evaluation Test Plan |
| SID | Standard Instrument Departure |
| SH | Super Hornet |
| SLAM | Standoff Land Attack Missile |
| SOP | Standard Operating Procedures |
| SOW | Statement of Work |
| SSC | Surface Surveillance and Control |
| SSOM | Solid State Oxygen Monitor |
| SSWG | Systems Safety Working Group |
| ST | Super Typhoon |
| STA | Shuttle Training Aircraft |
| STAR | Standard Arrival Route |
| Strike | Coordinating agency aboard the aircraft carrier |

|            | Tactical aviation test squadron at NAS Patuxent River, MD |
|            | Training organization at NAS Fallon |
| TAC        | Tactical Air Coordinator |
| TACAN      | Tactical Air Navigation |
| TACTS      | Tactical Aircraft Combat Training System |
| TAMPS      | Tactical Aircraft Mission Planning System |
| TANS       | Tactical Air Night Attack |
| TAS        | True Airspeed |
| TC         | Terrain Clearance; Test Conductor |
| TCM        | Technical Coordination Meeting |
| TEMP       | Test and Evaluation Master Plan |
| TF         | Terrain Following |
| TIFS       | Total Inflight Simulator |
| TLX        | Thin Line Explosive |
| TRAM       | Target Recognition and Attack Multi-Sensor |
| UFCD       | Up Front Control Display |
| USA        | United States of America |
| USMC       | United States Marine Corps |
| USN        | United States Navy |
| USNTPS     | U.S. Naval Test Pilot School |
| USO        | United Service Organizations |
| USSR       | Union of Soviet Socialist Republics |
| VFR        | Visual Flight Rules |
| VMC        | Visual Meteorological Conditions |
| VSI        | Vertical Speed Indicator |
| VTR        | Video Tape Recorder |
| WASEX      | War At Sea Exercise |
| WESTPAC    | Western Pacific |
| WG         | Working Group |
| WSD        | Weapons System Description |
| WSO        | Weapon System Officer |
| WST        | Weapon Systems Trainer |
| WTT        | Weapons Tactics Trainer |
| VX-5 / VX-9 | Operational Test squadron at NAS China Lake |
| XO         | Executive Officer |

## About The Author

David Maybury served as a Naval Flight Officer, Engineering Flight Test Officer and Aerospace Engineering Duty Officer. He flew attack aircraft from aircraft carriers, tested supersonic aircraft, provided engineering management for F/A-18 aircraft and led an aircraft structural repair detachment in Okinawa, Japan. David has master's and bachelor's degrees in aerospace engineering from the University of Colorado, is a 1990 graduate of the United States Naval Test Pilot School and possesses an Airline Transport Pilot license.